THE IRON HORSE

THE IRON HORSE

THE HISTORY AND DEVELOPMENT OF THE STEAM LOCOMOTIVE

JOHN WALTER

To Findlay, Georgia and Holly, who all like to ride on the train ...

Cover illustrations: front: a Dreyfuss-styled 4-6-4 Hudson of the New York Central Railroad departs the city in the late 1930s for Chicago, at the head of the *New 20th Century Limited*. *Back: top*: a drawing of a Dübs-made 4-4-0, published in 1892 in Jamieson's *A Text-Book on Steam and Steam Engines*; *below*: a French 4-6-2 Pacific of the Nord Railway after the modifications made by André Chapelon to raise power and efficiency to previously unknown heights.

First published 2016

The History Press
The Mill, Brimscombe Port
Stroud, Gloucestershire, GL5 2QG
www.thehistorypress.co.uk

British Library Cataloguing in Publication Data.
A catalogue record for this book is available from the British Library.

ISBN 978 0 7509 6716 7

Typesetting and origination by The History Press
Printed in China

CONTENTS

FOREWORD

This project began life in the 1990s as a guide to an exhibition. I worked for the British Engineerium in Hove, which was then not only one of the most important steam museums in the south of England but also a centre for conservation skills. There was a 'spare wall' between the exhibition hall and our workshop, and so I decided to create *The Iron Horse: development of the steam railway locomotive, 1804–1960* to fill the space.

Bachmann, Fleischmann and Märklin were all willing to donate model locomotives ranging from *Rocket* to *Big Boy*, allowing a 'little-to-large' progression by wheel notation to be created in a space just 40ft wide. Biographies of locomotive engineers were to grace the back wall, and we had excellent large-scale models to showcase. I persuaded John Snell, who was then managing the Romney, Hythe & Dymchurch Railway, to lend us one of his 'Miniature Pacifics' as a centrepiece, agreeing that we'd make our exhibition portable enough to take a trip to New Romney. I'd even gathered sponsorship to cover virtually all of our costs.

Then the problems began. For reasons well beyond my control, what I'd seen as an inexpensive space filler became increasingly sophisticated; the budget climbed alarmingly. Our backers understandably lost interest, and then the RH&DR locomotive I hoped to borrow had to be pressed into service after a 'traffic engine' failed in service.

Nothing more was ever done with the exhibition idea. I kept the draft of my handbook, updating it to become a small book. Then, in 2015, The History Press became involved through Michael Leventhal and Amy Rigg, the commissioning editors, and a decision was taken to publish.

The original text was deemed to be too short. The book had been conceived as a simple guide, and even though I was happy to double the length of the introductory material, we knew that the convoluted history of the locomotive engine cannot be summarised in a few thousand words. I was keen to avoid making 'yet another picture book', but balancing detail with broad-brush narrative can be difficult.

I decided that my approach should be episodic: concentrating on a handful of topics in some detail, but trying simultaneously to set them into a narrative context. A few biography panels would be retained, concentrating on lesser-known names. Of course, I knew that this would be met with howls of protest from the champions of Aspinall, Churchward, Fowler, Gooch, Gresley, Macintosh, Maunsell, Robinson, Stanier, Stirling and Webb – even excluding certain people from this list is potentially divisive! But many of these men have had detailed biographies of their own.

I have tried to be as even-handed as possible, and to place developments against an international backcloth. I realise that the first steps taken in North America are not discussed in minute detail, but I hope that coverage, judged overall, has universal appeal.

Dr Jonathan Minns (1938–2013), Director of the British Engineerium, supported the work in its early stages; Dr Richard Leonard very kindly gave permission to use photographs from his fascinating website; and the usual suspects – Alison, Adam, Nicky, Findlay, Georgia, Holly, and Jack the Dog – gave love, support … and cake.

John Walter, Portslade, 2016

No 317

PROLOGUE

The enthusiasm with which railways were embraced from the 1830s onwards created an ever-increasing need for locomotives that were faster and could haul greater loads than their predecessors. Carriages became progressively heavier in order to handle increasing demand, while freight tonnage – and the urgency with which it was wanted – also rose substantially. This created a need for two different classes of engine, which lasted for the remainder of what became known as the 'steam era'.

Once the initial fears had been overcome, passengers demanded to be moved more quickly; and the railways, realising that they had to compete with each other to be successful, were quite willing to provide faster services. Passenger engines, therefore, almost always had large-diameter driving wheels and, certainly after the 1850s, a leading bogie or swivelling single-axle truck to allow curves to be taken at high speed. However, though the large wheel allowed a greater length of line to be covered for each stroke of the piston, the tractive force available for pulling a load was reduced proportionately. In addition, as only one or two driving axles were used, little more than half the engine weight was carried on the driving wheels. Called the 'adhesion weight', this showed whether the engine could start a heavy load away from a station and climb gradients reliably. Some passenger locomotives used on hilly routes had quite small wheels, but these were still usually accompanied by a leading bogie or truck.

The most important goal of a goods or 'merchandise' engine, as it was called in the early days, was to pull as much as possible. Speeds did not need to be high, so small wheels were fitted. These engines did not need bogies or trucks, so all their weight could be carried on the driving wheels. Multiplying axles allowed locomotives to grow larger and more powerful, though lengthening the wheelbase could restrict the sharpness of the curves that could be negotiated and often promoted excessive wear on the wheel-tyres and even the inner face of the rails.

Some railways settled on compromise designs, often where perishable freight had to be moved quickly (e.g., fruit, fish or refrigerated meat), but these engines rarely had more than a leading truck prior to 1900. The 'mixed-traffic' engine only became popular during the twentieth century, combining medium-diameter wheels with as many driving axles as possible.

At the beginning of the Railway Age, the gauge, or distance between the tracks, was not settled. Though many lines kept to the 4ft 8½in recommended by the Stephensons, gauges of 5ft 6in, 6ft or even 7ft were also used. The Great Western Railway broad gauge (7ft 0¼in) had considerable stability advantages over the 4ft 8½in gauge, but the lobbying power of the standard-gauge railways prevailed and the last broad-gauge lines were converted in 1892. However, the meagre dimensions of the standard-gauge hamstrung British railway development in many ways – e.g. by restricting the dimensions of locomotive cylinders or the space allocated in passenger carriages.

As train loads grew progressively larger, the haulage ability of the railway locomotive was tested to its limit. The easiest method of increasing power was to enlarge the cylinders, extend the fire grate and lengthen the boiler. However, this increased engine weight considerably and increased

Typifying the Golden Age of British steam, A1 class Gresley Pacific (4-6-2) No. 4476 *Royal Lancer* of the London & North-Eastern Railway (LNER) pulls away from King's Cross station, London, in the period between the world wars. The locomotive was outshopped in May 1923, rebuilt to A3 standards in 1946 and was withdrawn in November 1963 as No. 60107.

the loading on each individual axle unless additional wheels were added. Thus the 2-4-0 two-cylinder passenger engine evolved into a three- or four-cylinder 4-6-2 and ultimately to a 4-8-4; the 0-4-0 freight engine became an 0-6-0, an 0-8-0 and – eventually – a colossal Soviet 4-14-4. Freight engines invariably had two cylinders.

But even this was not enough: among alternative solutions to the ever-increasing loads were higher working pressures, many differing types of compounding, extra cylinders, the introduction of enormous articulated engines weighing as much as 500 tons, and a prolonged investigation into thermal efficiency.

How a Locomotive Works and How it is Tested

The steam locomotive of the 1950s differed very little from Stephenson's *Planet* of 1830 as far as its basic characteristics were concerned.

Typical of the mid-Victorian era were the Midland Railway 4-4-0s designed by Samuel Johnson, which replaced the rigid frame 2-4-0 popular with the designers of the 1860s. Johnson designed the first of his bogie engines for the Great Eastern Railway, where he had briefly been locomotive superintendent, but they were built under the supervision of his successor. The first ten 'Midland Bogies' were made by Kitson & Company of Leeds to handle passenger traffic of the Leicester & Manchester and Settle & Carlisle sections of the Midland Railway. Dating from 1876, they handled loads of 150 tons on gradients of 1-in-100 at 35mph, burning 26–30lb coal per mile and evaporating 7–8lb water per pound of coal. Kitson engines were supplemented by thirty built in the Derby railway works in 1882–83. These had 6ft 9in coupled wheels, 18 × 26in cylinders inside the plate frames, and weighed 75 tons 12cwt in running order.

A drawing of the Johnson 4-4-0, from Daniel K. Clark, *The Steam Engine* (1891 edition).

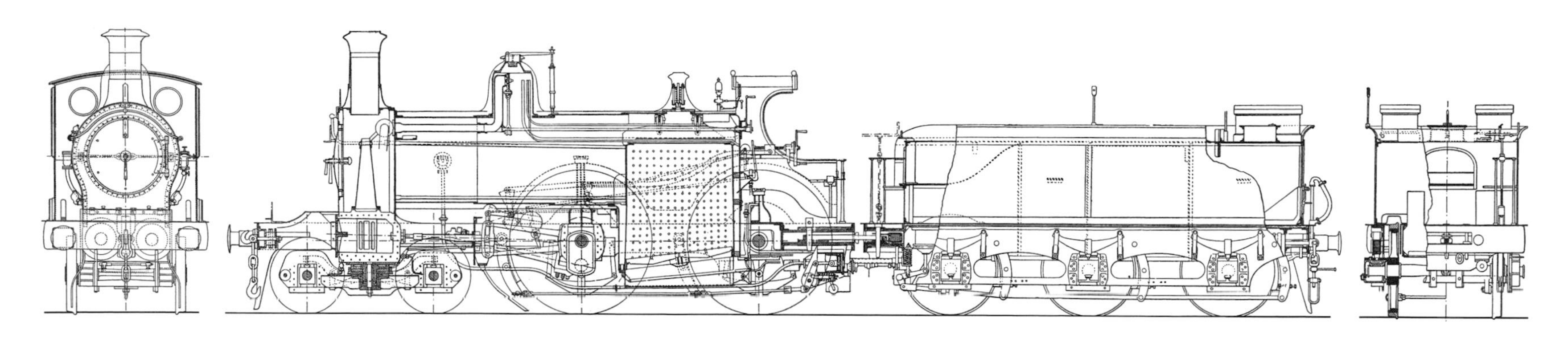

The Johnson 4-4-0s of the 1882–83 series were described in great detail by Daniel Kinnear Clark in his book *The Steam Engine*. The powerhouse of the engine was the boiler. This was basically a large-diameter cylinder, made of iron plates riveted together, closed at the rear by the firebox and at the front by the smokebox. Between the two ran 205 1⅛in-diameter fire tubes, nearly 11ft long, which connected the firebox with the front tube plate. This tube plate formed the back edge of the smokebox and provided a restraint for the water that filled the intervening space. The boiler had a capacity of 152ft³, usually comprising 114ft³ of water (about 710 gallons) and 38ft³ of steam space. Coal was introduced periodically to the firebox, through a specially hinged door, to burn with the assistance of a brick deflector arch. This was perfected in the 1860s – after much experimentation by Irishman Joseph Beattie among many others – to allow substitution of cheap coal for the expensive coke that had previously minimised smoke emissions. Hot smoke and gas rising from the fire passed from the firebox along the fire tubes and out into the chimney through the smokebox, helped by an air draught set up from fire to chimney.

The tubes radiated heat to the water in the boiler, until the temperature rose far enough for boiling to begin. Steam formed above the surface of the water, excessive pressure being prevented by the safety valve. The pressure in the boiler, in the case of the Johnson 4-4-0, was 140psi; at this juncture, the boiling point of water (which varies with pressure) was 356°F. When sufficient steam had been generated, the driver opened the regulator. This was a valve allowing steam to flow into the steam pipe and was usually placed in a dome on the top of the boiler to isolate the regulator from the worst effects of turbulence at the surface of the water. This prevented excessive moisture reaching the cylinders (called 'priming').

The steam pipe led forward inside the boiler casing, where it was surrounded by steam to prevent unnecessary loss of heat, until it entered the steam chest attached to the cylinder block. The block contained the driving piston and the valve mechanism. The valve admitted steam on alternate sides of the piston, pushing it backward and forward. The piston rod was connected to the driving axle by a sturdy strut – the connecting rod. If the cylinders were placed inside the frames (as in the case of the 4-4-0), the drive was to a crank formed in the axle itself; if the cylinders were outside the frames, crankpins in the centre disc of the wheels sufficed. By the time that steam reached the valve chests, its pressure had declined appreciably. It also contained a substantial quantity of moisture in suspension. Called 'saturated steam', this tended to condense in the cylinders and required the provision of cocks to drain it periodically.

The key to the operating process was provided by the valves. Flat-faced slide valves had become universal by 1870, operating in conjunction with two admission ports (one at each end of the cylinder) and a central exhaust port. The valve was connected with the crankshaft by a rod and an eccentric (often spelled 'excentric' in the nineteenth century), which is simply a disc with an offset hole. A strap or collar running around the periphery of the eccentric disc was attached to the valve rod. As the crankshaft rotated, the eccentric moved the valve rod backward and forward, alternately admitting and exhausting steam from the opposing sides of the piston.

Steam exhausting from the cylinder was led up a central blast pipe to emerge beneath the chimney. As the steam still had considerable pressure, its blast helped to drag smoke emanating from the fire tubes out the chimney. This in turn improved the flow of air through the grate (by creating a partial vacuum in the smokebox) and helped the fire to draw. Adjustable jumper tops or blast pipe rings were fitted to some engines to control the blast, which was sometimes so strong that small pieces of the fire were drawn through the fire tubes and ejected from the chimney with the smoke and steam.

Water had to be pumped back into the boiler from the tender tank, relying on two Gresham & Craven injectors. Patented in France on 8 May 1858 by the French engineer Baptiste-Jules-Henri-Jacques Giffard (1825–82), an injector can use live steam to force water into the boiler even though great pressure was being maintained. Injectors replaced manually operated force pumps, used on the earliest locomotives, and a selection of donkey pumps.

The most important feature of the railway locomotive, and often a key distinguishing feature, lay in the style and construction of the valve gear. Valve gear regulated the flow of steam into the cylinders, and, when necessary, could be altered by the driver to optimise performance. Perfected valve gear – there were many competing designs – allowed gabs, eccentrics, links or similar fittings to change the motion of the locomotive simply by redirecting the thrust of the piston, by way of the connecting rod, from its set position to the opposite side of the drive axle. Without altering the way in which steam was admitted to the cylinder, this allowed the direction of travel to be reversed.

By 1870, the system had been refined to ensure efficient use of steam by varying the point at which the admission of 'live' (full-pressure) steam to the cylinder ceased. This was called the 'point of cut-off'. Once a locomotive was running and the use of steam could be reduced, cut-off could be adjusted to use less live steam, completing the piston stroke by expansion of the steam in the cylinder.

The earliest forms of 'gab' valve gear (which could not be adjusted for cut off) had been superseded by the radial link gear designed in 1841–42 by William Williams and William Howe, then exploited by Robert Stephenson & Company of Newcastle upon Tyne from 1843 onward. This relied on a pair of eccentrics and a lifting link, operated by a reach rod in the cab, to re-set the valve spindle.

Stephenson gear (as it soon came to be called) was simple and efficient, but had its rivals. Even in the nineteenth century, Allan's straight link motion was popular, as it reduced vertical movement by shifting simultaneously both the straight link and the valve-rod die block; Joy's gear abandoned eccentrics altogether in favour of separate anchor, jack and vibrating links. However, as locomotives became progressively more powerful and cranks became larger, the space inside the frames was increasingly restricted. A major disadvantage of Stephenson gear was the need for two eccentrics per cylinder, and a move was eventually made towards valve gear that lacked eccentrics altogether.

By far the most popular of the 'non eccentric' valve gear (in Britain and Europe at least) was patented in Belgium by Friedrich Fischer in October 1844, but is best known by the name of its designer: a machinist named Egide Walschaerts (1820–1901). Poor Walschaerts was a mere mechanic and lacked the certificates and diplomas that would have allowed him to patent his invention. Though he was able to persuade his friend Fischer (and subsequently also Fischer's son) to act on his behalf, there is no evidence that he ever made money out of the valve gear; forty years after the first patent, Walschaerts was still working as a foreman mechanic. His contribution to the locomotive went unacknowledged for many years, particularly in Europe where Edmund Heusinger von Waldegg (1817–86) patented a similar system in Germany in 1849.

There has never been any evidence to show that either inventor was aware of his rival's design, but what is invariably known as 'Walschaerts' gear' in English-speaking countries is still 'Heusinger's gear' in central Europe. It relied on a return crank attached to the crankpin, which drove a slotted link by way of an intermediate rod. William Mason introduced Walschaerts' gear to the USA in 1876, on the Mason Fairlies (colloquially known as 'Mason Bogies'), and the first locomotive with Walschaerts' gear to be made in Britain was an 0-4-4 'Single Fairlie' built by the Avonside Engine Company of Bristol in 1878. This machine ran for a few years on the Swindon, Marlborough & Andover Railway, beginning in 1882, but soon proved to be exceptionally uneconomical and so Walschaerts' gear was rarely seen elsewhere in Britain prior to 1914 – but it was in Ireland. Many Irish engines were supplied by the early British champion of the mechanism: Beyer, Peacock & Company of Manchester.

Ultimately, in the USA, a valve gear protected by US Patent No. 1008405, granted on 14 November 1911 to Abner DeHaven Baker (1861–1953) of Swanton, Ohio, gained ground on its rivals. Though the Brooks Locomotive Works was one of a handful of manufacturers to promote the use of Stephenson gear (and attempts were made from time to time to establish the use of Walschaerts'), Baker gear promoted by the Pilliod Company of Swanton made great headway and, by the 1920s, had become pre-eminent in North America. There is little doubt that Baker gear outperforms Walschaerts' gear in important respects. And yet, strangely, it is virtually unknown in Europe … perhaps another testimony to the widespread belief in Europe (and in Britain in particular) that very little of note was achieved by North American locomotive engineers!

When the first locomotives appeared at the beginning of the nineteenth century, understanding of thermo- and fluid dynamics was in its infancy.

A 1977 vintage painting, '230G353' by Claude Yvel, showing Walschaerts' valve gear in detail. (Courtesy of Galerie de Luxembourg, Paris)

The idea of comparing power to that of a horse had been established, and practical tests – how many wagons could be drawn, how much water could be raised in a given time – were undertaken regularly. The work of English engineer John Smeaton (1724–92) was particularly valuable as it established yardsticks such as 'Duty' and 'Great Product', which allowed proper comparisons to be made. Smeaton underestimated the value of a single horsepower (he is said to have based his work on pit ponies!), allowing James Watt, employing a 'brewery dray horse', to subsequently raise the value to its currently accepted level.

The first steps towards standardised testing had been made. The competitive nature of trials, most notably among Cornish mine captains, soon promoted improvements. Gradually, the power and efficiency of what was essentially an inefficient machine grew appreciably. In the 1780s, James Watt devised a simple indicator to show how steam pressure within a cylinder varied during the piston stroke. In 1796, Watt's employee, John Southern, fitted Watt's indicator with a wooden tablet, attached to a suitable reciprocating part of the drive mechanism, which moved laterally as the pointer attached to the piston moved vertically. The result was the 'indicator diagram', effectively a closed loop, showing how pressure of the steam within the cylinder varied from entry to exhaust.

The indicator diagram was a revelation as it allowed the internal working of the steam engine to be investigated in detail. It showed the economies that could be made by 'expansive working', when full-pressure steam entering the cylinder from the boiler was cut off during the stroke and allowed simply to expand until the exhaust valve opened. The indicator also showed that expansive working was most beneficial if boiler pressures were higher than those customarily used by Boulton & Watt. The ideals of Cugnot, Oliver Evans, Trevithick and others had been vindicated.

Unfortunately, the unwieldy Watt indicator was unsuitable for applications where vibration or rapid movement were expected. Sometime in the 1820s, Scotsman John McNaught developed a portable indicator in which an exchangeable 'power spring' inside a cylindrical body resisted the upward movement of a piston while a spring-loaded drum – initially co-axial, later offset – was turned around and back by a cord attached to a crosshead or suitable link. A pointer attached to the piston traced the diagram on a sheet of paper attached circumferentially to the drum.

The McNaught indicator was made in sufficient numbers to be influential. Though severe vibrations could give an unsatisfactory trace, the instruments were used regularly with land and marine engines. The railway locomotive, however, presented a different challenge. Irregularities of the track and the tendency of the earliest two-cylinder locomotives to oscillate laterally were enough to prevent satisfactory use of McNaught and similar indicators in all but the most ideal circumstances.

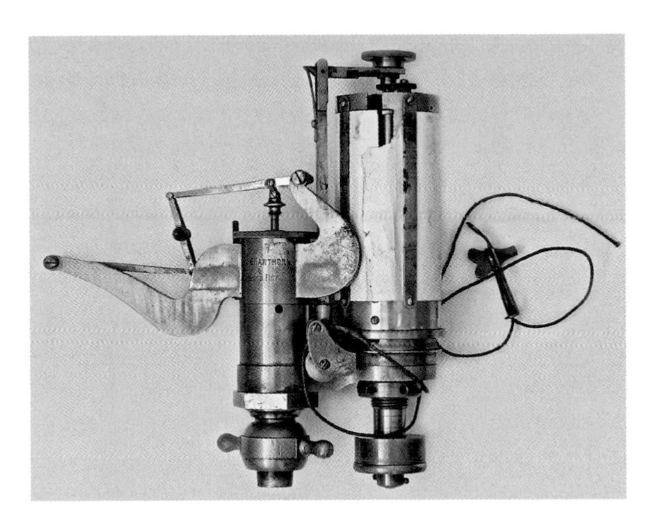

A Richards indicator fitted with a Richardson-patent continuous recording mechanism. This particular instrument was used by the locomotive department of the engineers R. & W. Hawthorn.

The first person to provide an indicator suitable for the railway locomotive was Daniel Gooch, who wished to investigate the performance of those running on the Great Western Railway during the 'Battle of the Gauges', when he was determined to prove the merits of the 7ft 0¼in broad gauge over the standard 4ft 8½in. Gooch built a dynamometer carriage (the first of its type) that recorded the pull the locomotive could exert on the draw-bar connecting its tender and the train. This 'tractive effort' was not only a clue to the maximum load that could be hauled but also, in conjunction with the indicator diagram, gave a clue to mechanical efficiency: friction, heat loss and other factors usually conspired to reduce usable horsepower to about 80–85 per cent of 'indicated horsepower', and friction between the wheel types and the rail then reduced power considerably to what was effectively 'draw-bar horsepower'.

Gooch's indicator recorded the pressure/time trace in a way that was difficult to interpret compared with the clarity of the McNaught type. However, it worked reliably under circumstances where the latter did not. The advent of the first high-speed stationary engine, the Porter-Allen design of 1860, presented engineers with a problem; running speeds in excess of 300rpm, even if largely vibration free (which few of the earliest installations were!), was too much for a McNaught. An answer was provided by Charles

B. Richards (1833–1919), a freelance engineer of Hartford, Connecticut, to whom British Patent 1450/62 of 1862 and US Patent 37980 of 24 March 1863 were granted to protect a compact derivation of the McNaught. This allied short springs with a pantograph-type amplifier to enlarge the movement of the piston at the trace-point by a factor of four. This allowed a stiff spring to be used, accepting that even high pressures would give minimal compression, yet still produce a diagram large enough to be analysed effectually. At a stroke, the problems of vibration had been, if not conquered, at least minimised.

Elliott Brothers of London, makers of optical and scientific instruments, began to make Richards indicators in quantity in 1863. When Porter returned to buy several of them to use during preparatory work on the Centennial Exhibition in Philadelphia (1876), output was approaching 10,000.

The Tabor Steam Engine Indicator (*c.* 1889), written by George Barrus on behalf of the Ashcroft Manufacturing Company of Bridgeport, Connecticut, contains impressive testimonials and a list of more than 300 purchasers of Tabor indicators. By far the greatest number of entries refers to engineering businesses, but among the public utilities and shipping companies were the Illinois Central, Mexican Central, Norfolk & Western, Pennsylvania and Union Pacific railroads.

By the end of the nineteenth century, with protection conferred by the Richards patent long gone, so many other manufacturers were making indicators that the total was approaching 100,000. The lightweight Crosby design, made in Boston, Massachusetts, was particularly influential once piston-speeds of railway locomotives (and those of internal-combustion engines) began to climb.

Indicators allowed data to be collected from engines 'on the move'. The wooden shelter on the buffer beam of this LB&SCR 4-4-0 No. 317 provided the operators with at least some protection from the elements.

A qualified engineer working in the USA in 1900 would undoubtedly have been able to buy an indicator with the equivalent of a week's wages. In Britain, where lack of competition ensured that indicators were much more expensive, a skilled engineering worker earned an average of only £2 5*s* (£2.25) per week in 1901, and a surveyor or engineer received £6 8*s* (£6.40); even for a qualified engineer, therefore, a £15 indicator would cost more than two weeks' wages.

More information about the engine indicator, how it works, and the way in which results are analysed can be found at www.archivingindustry.com/Indicator.

If the indicator could show how the engine was performing, at least theoretically, *tractive effort* ('TE') was often quoted as a guide to hauling power. It is calculated by the formula:

$$\frac{0.85 \times D^2 \times S \times P \times n}{2W}$$

'D' is the diameter of the cylinder in inches; 'S' is the length of the piston stroke in inches; 'P' is boiler pressure in psi; 'n' is the number of cylinders; and 'W' is the diameter of the driving wheels in inches. To allow for friction, lost motion and other limitations, efficiency is assumed to be 85 per cent.

Tractive effort, though a useful yardstick, does not tell the whole story. For example, it will not be clear if the boiler can supply enough steam to run at speed, and there is no clue to the 'adhesion factor' (the proportion of locomotive weight carried on the driving wheels). If the adhesion factor is low, haulage capacity is likely to be similarly restricted; consequently, an 0-10-0 will be a far superior haulier than a 4-2-4 even assuming the driving wheels to be the same size.

But tractive effort does show how the power of the steam locomotive has increased. *Locomotion No. 1* (1825) of the Stockton & Darlington Railway had a calculated TE of 1,920lb compared with 16,270lb for 4-4-0 *No. 999* (1893) of the New York Central & Hudson River Railroad, 31,625lb for 4-6-0 No. 5073 *Caerphilly Castle* (1923) of the Great Western Railway, and a staggering 84,980lb for the J-Class 4-8-4 (1941) of the Norfolk & Western Railroad. However, *Locomotion* weighed merely 8 tons 8cwt empty; a J-Class engine with its enormous tender weighed 437 short tons! A calculation of tractive effort per ton actually favours *Locomotion*, unless the tender of the Norfolk & Western engine is ignored.

Wheel Arrangements

No attempt to summarise the history of the railway locomotive would be possible without an understanding of the conflicting systems of classification used to categorise them. In the English-speaking world, locomotives are now customarily classed in accordance with a system devised in 1900 by Frederick Methven Whyte (1865–1941). However, Whyte's annotation sometimes makes it difficult to distinguish between driven and carrying wheels; or, alternatively, axles carried in bogies, trucks or rigidly in the frame.

At the time of its introduction in the USA, the Whyte classification was capable of satisfying virtually all 'main-line' applications, as the period of experimentation – at least so far as North America was concerned – was almost over. In Europe, however, the situation was less clear; Fairlie, Mallet, Garratt and comparable articulated machines could usually be accommodated in the Whyte scheme, but the oddities of railway history were often much more difficult to categorise. A system popular in continental Europe, which counted the wheels per side instead of in total and used letters to identify driven wheels, had great advantages. Unfortunately, it was rarely used in the Anglo-American world even though Oliver Bulleid of the Southern Railway tried to introduce it to Britain with his Merchant Navy-class Pacifics (4-6-2 according to Whyte, or 2C1 to Bulleid).

A challenge to the Whyte system was also posed by locomotives such as the early Webb compounds running on the London & North Western Railway, which had axles driven independently. Had coupling rods been fitted, the locomotives would have been readily classifiable as 2-4-0, 2-4-2, 4-4-0 or 2-6-0. Owing to the quirky drive system, they are often classified as '2-2-2-0', '2-2-2-2', '4-2-2-0' and '2-2-4-0', but this does not give an immediate indication of the position of the driving wheels.

Difficulties will also be encountered with the small number of engines driven by a single cylinder-cluster, but with the driven wheels split into groups. The Hagans patent engines, made in small numbers prior to 1914, had the rear group of axles (generally no more than two) on a special pivoting sub-frame. Though perhaps qualifying as 0-8-0 or 0-10-0, cursory observation is more likely to judge 0-4-4-0 or 0-6-4-0 to be realistic.

Unlike the Mallets, however, the drive units of the Hagans engines were originally coupled together, but not by a single rigid rod. The same is true of a class of French 2-10-0 machines, built for the PLM in the 1930s, which have an internal coupling between the third and fourth axles.

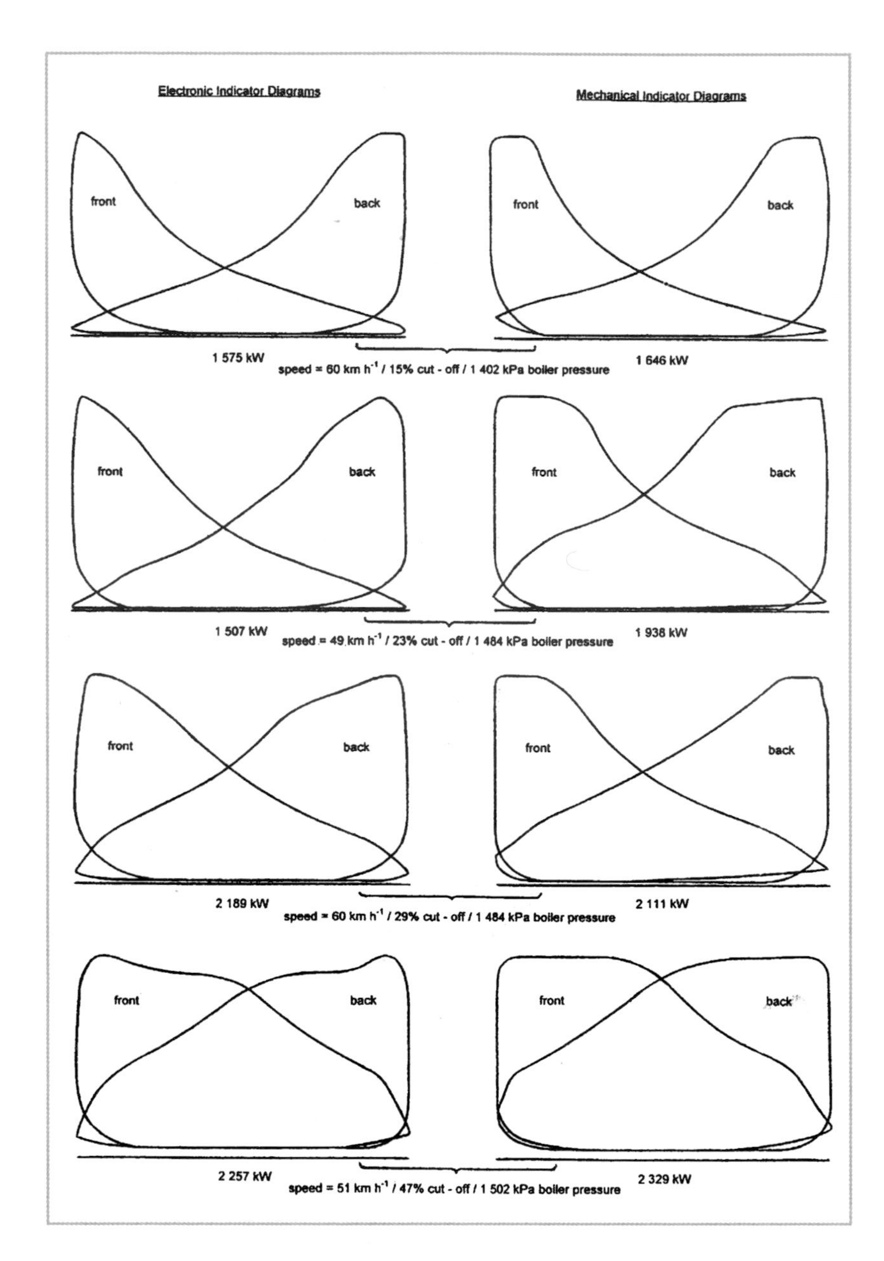

A series of typical indicator diagrams taken from a South African Railways Class 26 locomotive during an experiment to determine the efficiency of the mechanical indicator. Unfortunately, peculiarities of the valve gear ensured that the individual traces of the indicator and the electronic recorder could never be the same, though the general similarity is obvious.

The European alphanumeric system gives a more helpful picture of the construction of the London and North-Western Railway split-drive compound locomotives – e.g., 1AA1 for 2-2-2-2 – but it is regrettable that no system was ever devised to distinguish types of carrying axle, which could be fixed in the frame, supported in trucks or mounted in bogies. Among the most extreme of the earliest freaks were two locomotives made for the Great Western Railway by R. & W. Hawthorn of Newcastle upon Tyne, to the orders of the great engineer Isambard Kingdom Brunel.

Patented in 1836 by Thomas Harrison, they had cylinders and running gear on one frame and the boiler on another, relying on ball-joint connectors to convey steam from the boiler to the cylinders. *Hurricane* had a single driving axle with 10ft-diameter wheels, the largest known to have been fitted to a railway locomotive. Its theoretical notation was 2-2-2+6, as the boiler was carried on a six-wheel carriage.

Thunderer was essentially similar, but the drive unit was an 0-4-0 with 6ft-diameter coupled wheels. Gearing was used to increase the speed of rotation. Harrison's goal was apparently to keep axle loading as low as possible, but the lack of adhesive weight doomed the locomotives to failure.

Another strange locomotive is said to have been supplied by Robert Stephenson & Company of Newcastle upon Tyne to a mine railway in Antofagasta, Chile, in 1884. This was apparently a Webb-compound tank engine with the extraordinary wheel arrangement of 4-2-4-2. A few machines were made with carrying axles between the driving axles, usually to correct bad distribution of weight, and there were even some with auxiliary or 'selectable' wheels to increase tractive power when required. One locomotive of this type, patented by Krauss & Co. of München in 1896, was tested by the Bavarian state railway in 1897; a larger version was proffered in 1902. The first machine was a 4-2-2 convertible to 4-2-2-2 (the auxiliary drive axle lay behind the bogie); the later example was a 4-4-2 convertible to a 4-2-4-2, with the supplementary axle between the bogie wheels.

Another challenge to the Whyte notation is provided by the wide variety of articulated locomotives that have been made. The Garratt is easy to categorise as it is essentially two locomotives joined by a single central boiler: if it is essentially two ten-wheelers (4-6-0) joined back to back, then the notation would clearly be 4-6-0+0-6-4. The standard Fairlie locomotives are clearly two self-contained bogies beneath a single boiler/frame unit, and are more accurately classed as 0-4-0+0-4-0 than the more normally encountered '0-4-4-0'. A more obvious problem concerns the

Posed in front of an outstanding example of ornamental brickwork in an industrial application, 4-4-0 No. 318 of the Lancashire & Yorkshire Railway was built in 1894 in the company's Horwich Works to run express passenger trains. The diameter of the large driving wheels was 7ft 3in and the two inside cylinders measured 18 × 26in.

This curious 4-2-2, No. 1400, was made by Krauss for the Bavarian state railways in 1896. The two-cylinder compound had a booster, effectively a small single-cylinder auxiliary engine, which could be engaged when required. The cylinder lay directly below the main cylinders, with its driving axle (shown raised in the illustration) behind the rear bogie wheels. The engine ran trials for at least two years, but the additional complexity was not justified by results.

One of twenty-two 0-4=4-0 Mallets made by SACM for the Chemins de Fer de la Corse (Corsican railway), apparently between 1893 and 1932. The last example was withdrawn from service in 1954.

This Climax-geared locomotive was delivered to the Hillcrest Lumber Company of Vancouver, BC, in March 1928 – the penultimate of its type to be built. (Mount Rainier Scenic Railroad and Museum, Mineral, Washington State, USA)

Mallets, which have one power unit integral with the boiler/frame unit and another pivoted at the front of the rear unit to enable what would otherwise have been too long a rigid wheelbase to follow a curve. The original Mallets were compounds, which meant that high-pressure steam (usually the rearmost cylinders) was fed to the low-pressure cylinders on the front power unit. However, the locomotives give the appearance of a rigid frame (which the Pennsylvania Railroad duplex classes actually were) and so are normally listed as 2-6-6-2 or 4-8-8-2. There is a case for using 2-6=6-2 or 4-8=8-4 to show instantly that the power units are separate, but not in the dissociated way of the Garratts ('2-6-0+0-6-2').

The Shays, the Climaxes and Heislers all have multiple all-wheel drive bogies driven by a single drive-train, though the construction and position of the cylinders vary. The locomotives are probably best listed as 0-4-4-0, though 0-4-0:0-4-0 could be a better indicator (in these cases, the continental Europe system would be much better, as they would simply be 'B:B' or 'B:B:B' for the two- and three-bogie versions respectively).

Perhaps the craziest of all locomotives was a unique two-boiler machine built in Belgium in the 1930s that had two four-axle power bogies and a multi-axle central frame. Its Whyte notation would have been a scarcely believable 0-6-2-2-4-2-4-2-2-6-0, or, in European terms, 0C11B1B11C0 ...!

Although I have retained the Whyte notation on the grounds that it is the most familiar to English-speaking readers, there is considerable merit in the continental method of noting a Pacific (4-6-2) as '2C1', as this immediately enumerates the driven wheels.

The current UIC system, supposedly to be used internationally, is essentially similar to the continental method except that it adds suffixes that indicate, for example, a superheater and the type of tender. This is generally reckoned to be unnecessarily complicated by non-technical enthusiasts, as the presence of a superheater, for example, is not always obvious externally. In addition, the quirky written notation of the UIC system can easily confuse the uninitiated.

1

THE FIRST STEPS

The Railway Locomotive, 1803–46

The first attempt to propel a vehicle by steam is generally credited to Salomon de Caus, who, in 1615, described and illustrated the expulsion of water from an orifice in an otherwise closed vessel that when heated could move a light wheeled carriage. Frenchman Denys Papin, however, is due the credit for the first piston-in-cylinder engine – albeit in an extremely primitive form. He also invented a means of deducing chamber pressure by sliding a weight along a bar, without realising that in this lay the basis of an effectual safety valve. Papin is said to have built a model road locomotive propelled by his piston-and-cylinder engine in 1698, but nothing else of note was done until 1757–58, when James Watt began research into the steam engine at the request of Dr James Robison of Glasgow University. His first model, never finished, was apparently of a steam-powered road carriage.

The earliest full-size locomotive engine, the *fardier à vapeur*, was created by the French artillery captain Nicolas-Joseph Cugnot (1725–1804), a small-scale prototype built in 1769 being followed rapidly by a vehicle large enough to be considered as an artillery tractor. Unfortunately, the three-wheeler was unstable and overturned on a bend. The project was then abandoned. Many a joke has since been had at poor Cugnot's expense, without recognising not only that he was working at a time when a 'steam engine' in France was customarily the wheezing, ponderous Newcomen atmospheric design, but also that he had used two cylinders.

In 1780, another Frenchman, Thomas-Charles-Auguste Dallery of Amiens, is said to have made a road locomotive fitted with a water tube boiler. The machine had a single central two-wheeled axle with small balancing wheels at each end, but was no more successful than Cugnot's had been.

Watt, Murdock, Symington, Evans and others experimented throughout the 1780s, but not until Richard Trevithick made the first of his road locomotives in 1801 was real progress made. Trevithick's machine was destroyed by fire after making a trial trip or two and a replacement, demonstrated in London in 1803, was not successful enough to arouse interest. By this time, however, Richard Lovell Edgeworth, contributing to *Nicholson's Journal of Natural Philosophy, Chemistry and the Arts* in 1802, had become one of the first to suggest the application of steam power to wagon ways. He had in mind the use of stationary engines, though the first successful application was not made until 17 May 1809 on the Bewicke Main wagon way.

Richard Trevithick, born on 13 April 1771 in Illogan, Cornwall, is simultaneously regarded as the father and the *enfant terrible* of steam locomotive history. The son of a Cornish mine captain, Trevithick had developed a high-pressure stationary engine that had brought conflict with Boulton & Watt. A series of inventions followed, culminating in the road carriage from which the first railway locomotive was developed. A demonstration of 0-4-0 *Catch Me Who Can* on a circular track in London in 1808, the first 'public railway' and also the first locomotive to rely on drive directly to the wheels, was unable to convince the public of the merit of railways. Worn down by ill-health, bankruptcy and a succession of controversial (if often percipient) engineering projects, Trevithick lost interest. In 1816, he quit England for the Cerro de Pasco silver mines in Peru, where one of his high-pressure stationary engines had replaced an asthmatic Boulton & Watt some years earlier. Trevithick remained in South America – still pursuing hopeless ideals – until he was recognised on Cartagena quayside, penniless and at his wits' end, by Robert Stephenson. The latter paid part of Trevithick's passage back to Falmouth, to be reunited with his wife after an eleven-year absence. Projects including an early form of storage heater and a reaction turbine came to nothing and Trevithick, combative to the end, died penniless and alone in the Bull Hotel, Dartford, Kent, on 22 April 1833. The funeral expenses were paid by J. & E. Hall of Dartford (for whom the turbine was being developed), but Trevithick's grave was unmarked. History has been kinder to Trevithick than his contemporaries ever were. He is now credited with the invention of the blast pipe, the fusible plug and even the multi-tube boiler. Trevithick married Jane Harvey, of the renowned engine-building family Harvey of Hale, and had several children. His son Francis (1812–77), one-time locomotive superintendent of the London & North Western Railway, wrote his father's biography; and other members of the Trevithick family helped to spread railways across the world.

The need to construct railways in all kinds of terrain, to meet differing operating criteria, has created a wide range of locomotive equipment. Among the earliest problems was the inability of wooden and cast-iron rails to bear the weight of locomotives that were powerful enough to be useful; another important limitation was the grip to be expected between smooth iron wheels riding on smooth surface iron rails, which would obviously limit the loads that could be hauled before slipping began. Whether this was the most effectual method of traction – indeed, even permissible – puzzled many early experimenters, until William Hedley built a mechanically powered test cart and proved that adhesion was sufficient by itself to be acceptable.

The earliest locomotives relied on beams, rods and levers to transmit power to the driving wheels. The first efforts (such as Trevithick's) had only one cylinder, though the boiler pressure was higher than that used by Boulton & Watt's stationary engines. Spent steam was exhausted directly into the atmosphere instead of a condenser. Relying on a single cylinder required a heavy flywheel to smooth out the intermittent power supply, but a glance at the Trevithick model will show that the clumsy flywheel was a source of danger.

The Coalbrookdale Company was the first to build a horizontal-cylinder engine, designed by Trevithick to run on a wagon way. Work had begun by August 1802, but was stopped when the engine was 'first started up' (possibly owing to a boiler failure) and was then abandoned.

In 1804, Trevithick completed a locomotive engine for the Pen-y-Darren Ironworks near Merthyr Tydfil. The owner of the works, Samuel Homfray, had so much faith in the engine that he bet 500 guineas with Anthony Hill of the rival Plymouth Ironworks that 10 tons of iron could be hauled on the 4ft 6in-gauge wagon way from Pen-y-Darren to Abercynon. The wagon-way locomotive was tested successfully on 13 February 1804. It had a single 8¼ × 54in cylinder and ran at about forty strokes per minute, giving a speed of 4mph. The exhaust steam was turned from the cylinder up the chimney, forming a primitive blast pipe: an invention that was later claimed on behalf of Stephenson, Hackworth and many others. Steam was distributed to the cylinder through a four-way rotary valve, which was operated by tappets on a rod struck by a lug on the crosshead. The only safety feature appears to have been a fusible plug, which may also be due to Trevithick. The plug melted if the boiler water evaporated and the temperature of the boiler wall rose too far.

When the wager run took place on 21 February 1804, the engine successfully pulled five wagons, 10 tons of iron and about seventy joyriders from Pen-y-Darren to Abercynon. The 9-mile journey took a little over four hours, on 2cwt of coal, as undergrowth had to be cleared from the path of the train. Maximum speed had been about 5mph. On the return journey, however, a bolt holding the axle to the boiler barrel came away and the boiler water escaped. The locomotive made several more trips, hauling as much as 25 tons of iron, but was too heavy for the plate way and was eventually relegated to stationary duties. Yet several other Trevithick engines were made, including *Catch Me Who Can*, demonstrated on a circular track in London in 1808. This locomotive had a single vertical cylinder buried in the rear of the boiler, driving on to the rear wheels through rod and crank drive.

The earliest commercially successful locomotive was undoubtedly the rack-and-pinion pattern based on an English patent granted in April 1811 to John Blenkinsop, manager of the Middleton Colliery in Leeds. Though the railway did not last long in its original rack-drive form, it was the prototype of all later rack-and-pinion systems. The first two engines were made in a foundry in Leeds owned by Fenton, Murray & Wood, detail probably being due to Matthew Murray. The first engine entered service on 15 August 1812. It weighed about 5 tons and was powered by two vertical single-acting cylinders (said to have had a 9in bore and a 22in stroke) that drove a toothed pinion revolving between the carrying wheels. Suitable teeth were cast into the outside surface of the rails. The engines could haul 90 tons on level track, and even succeeded in taking a 15-ton load up a slope. Several of them had been made by 1815.

Ascribed to Richard Trevithick, this model is believed to date from about 1806. The single cylinder, placed vertically, mirrors the construction of *Catch Me Who Can*, which ran on a track in London throughout the summer of 1808.

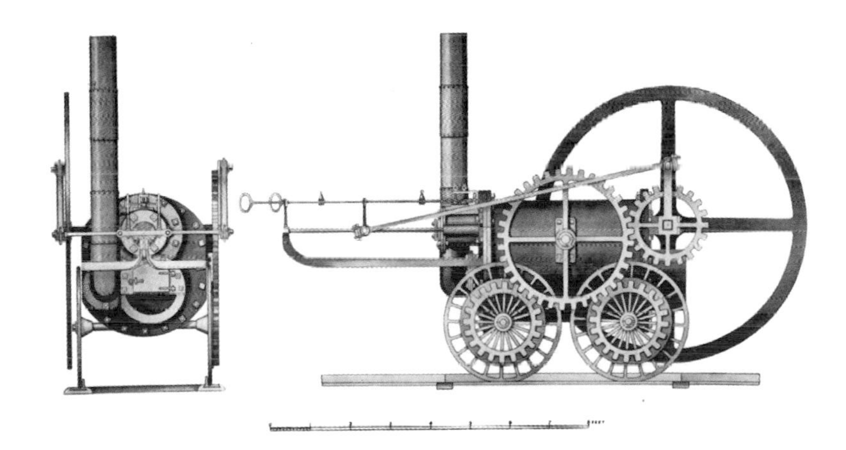

Trevithick's Pen-y-Darren locomotive of 1804 could be identified by the position of the chimney and cylinder. It was the first to run successfully, but proved to be too heavy for the track and had a short working life on the plate way. (Courtesy of the Science Museum; Crown Copyright)

The third Blenkinsop & Murray engine to run at Middleton Colliery had double-acting cylinders (its predecessors were rebuilt to this form), and eccentrics operated the valves instead of the original tappet gear. A fourth Middleton locomotive was fitted with a wooden condenser cistern above the boiler to receive steam exhausted from the cylinders, which otherwise frightened horses. Blenkinsop locomotives survived on the Middleton railway until 1834, when a brief return to horse traction was made before edge rails were laid for adhesion-type locomotives.

At this stage of railway history, the best method of propulsion was still to be discovered. On 30 December 1812, a patent was granted to William and Edward Chapman to protect a vertical-cylinder chain haulage locomotive engine, the reduction of pressure on the track arising from the use of multiple axles, and the mechanical equalisation of load on

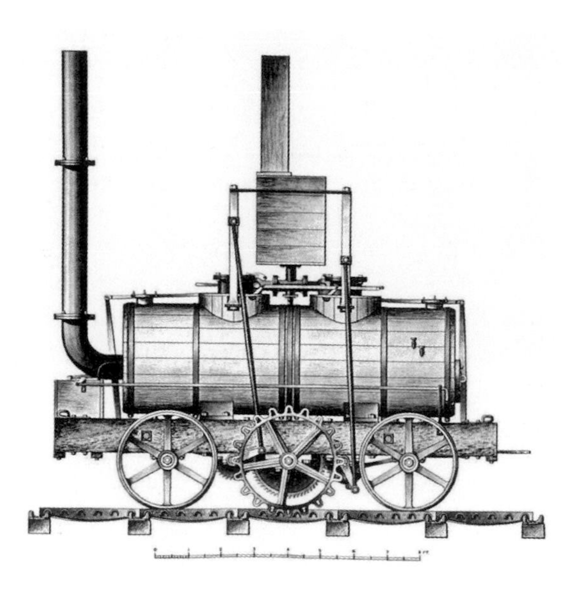

Made in Leeds by Fenton, Murray & Wood, Blenkinsop's 2-2-2 *Salamanca* was the first locomotive to haul useful loads on a railway. The cog-and-peg system allowed surprisingly heavy loads to be drawn up shallow gradients. (Courtesy of the Science Museum; Crown Copyright)

the wheels by mounting them in pairs on swivelling trucks or bogies – an idea subsequently widely claimed as 'novel' in Britain and the USA. The oldest Chapman engine, possibly a four-wheeler, was built prior to February 1813 by Phineas Crowther in the Ouseburn Foundry, Newcastle upon Tyne, but was soon reconstructed with an additional axle. The four-wheel wood-frame truck was attached to the underside of the boiler by a spherical pivot. One six-wheel example subsequently ran on the 4ft 5in-gauge tramway in Heaton Colliery. It had two 8 × 24in cylinders and a cast-iron return-flue boiler pressed to about 60psi. Power was taken from the cylinders by side levers driving chain wheels, guide pulleys and binding wheels. It was rebuilt extensively as a geared-adhesion type 0-6-0 in about 1817. The pivoting four-wheel truck was apparently retained, some play being allowed in the gears to accommodate lateral play.

In May 1813, Chapman & Buddle produced a promotional pamphlet containing drawings of an improved eight-wheel chain drive engine. An engine of this type may have been tried in Hetton Colliery in October 1813, but no other Chapman pattern engine is known to have been used until, on 21 December 1814, one built by Phineas Crowther pulled eighteen wagons (a load of 50–60 tons) up a 1-in-115 incline in Lambton Colliery. An account in the *Newcastle Chronicle* for 24 December confirms that the engine was an eight-wheeler running on two four-wheel trucks.

Not surprisingly, Chapman's chain-drive haulage system was not successful enough to survive against simpler methods of construction and disappeared rapidly. The development of the adhesion engine, relying on weight and friction between its wheels and rails to pull loads, is generally credited to William Hedley, viewer of Wylam Colliery. As a patent granted to Hedley in 1813 mentions rope or chain haulage, however, the design of an eight-wheel locomotive run in the colliery in 1814–15 is open to doubt.

Writing in 1836, Hedley claimed to have built his first engine (a four-wheeler) in March 1814 after experimenting with a man-power 'adhesion truck'. He also claimed that the engines – there were eventually several of them – were too heavy for the track and had to be rebuilt as eight-wheelers, before reverting to 0-4-0 when edge rails were laid in 1828.

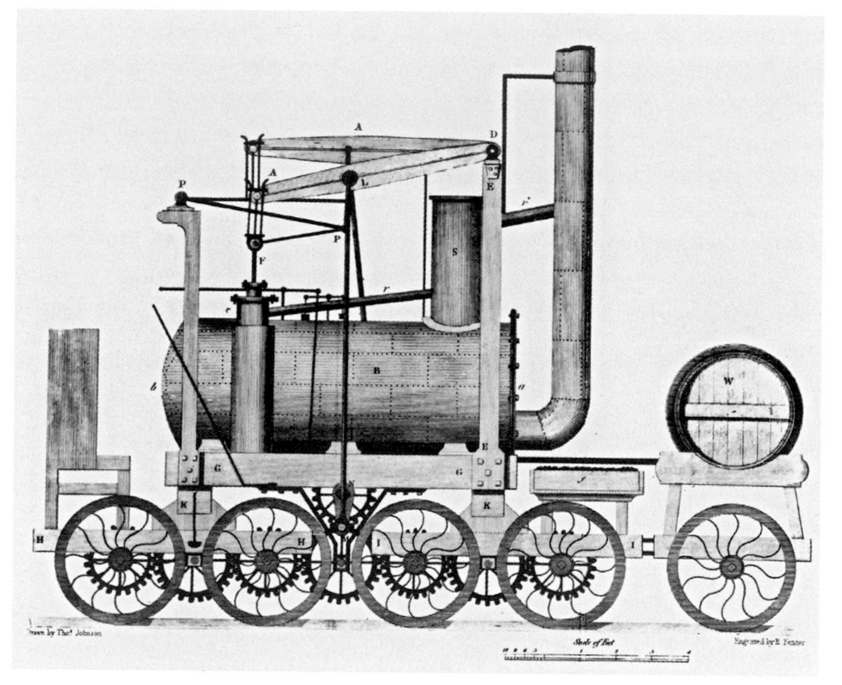

One of the locomotive engines built at Wylam Colliery to the designs of William Hedley. Note that the four axles were connected by gearing. After serving for a few years, improvements in the permanent way (and the introduction of superior designs) allowed the machines to be rebuilt in 0-4-0 form with side rods connecting the wheels.

Two surviving locomotives are credited to Hedley, blacksmith Timothy Hackworth and wheelwright Jonathan Forster: *Puffing Billy* in the Science Museum in London and *Wylam Dilly* in the Royal Scottish Museum in Edinburgh. Possibly made in Newcastle by Phineas Crowther, the former weighs about 8 tons 6cwt in running order. Both engines now have four wheels and the Fremantle design of parallel motion. However, though patented in 1803, Fremantle gear did not become commonplace until patent protection expired in 1817. This and other details suggest their current appearance, arising from rebuilding in 1828–29, hides their original form. Consequently, the influence of Hedley on the development of the locomotive engine is very difficult to gauge.

Two cylinders driving chains or gears through half-beams were used by almost everyone – William Losh and George Stephenson included – until improved forms of drive were developed. A bell-crank lever was fitted to Stephenson's *Experiment* of *c.* 1827, and a dummy crankshaft appeared in some of Hackworth's earliest designs. Finally, Hackworth fitted the piston rod directly to the wheels of *Royal George* (even though a parallel motion was retained) and coupled the three axles together. Stephenson's *Lancashire Witch*, built for the Bolton & Leigh Railway in 1828, was the first to embody inclined cylinders and a crosshead to support the piston rod.

The advent of springs had much to do with the relocation of the cylinders. Combining springs and vertical cylinders was soon shown to be unwise, as the engine tended to lift from the track each time the pistons thrust downward. This could be so dangerous at speed that the cylinders were moved until they were placed diagonally, but even this proved to be unsatisfactory and they were soon made nearly horizontal.

Other innovations included the provision of better valves (the conventional 'D' type was patented by Matthew Murray in 1802), the introduction of multi-tube boilers, and a rudimentary appreciation of the value of a blast pipe.

The First Public Railways

The opening of the Stockton & Darlington Railway in 1825 is now regarded as one of the most important events in early railway history. However, the earliest engines, exemplified by what was eventually known as *Locomotion No. 1*, were still derived from the beam engine and had vertical cylinders buried in the boiler. The first machine was ordered from Robert Stephenson & Company in September 1824, at a cost of £500, and delivered in September 1825. The prototype of four supplied in 1825–26, it was built to a specification supplied by George Stephenson, consultant engineer to the S&DR. It was basically a Table engine, with two 9 × 24in cylinders, axles connected by rods, valves driven by an eccentric and (originally) a single-flue boiler pressed to about 50psi. *Locomotion No. 1* and its four-wheel tender weighed 10 tons 2cwt in working order, and could pull loads of 55–75 tons at 5mph. Maximum speed was about 10mph.

In 1829, the directors of the Liverpool & Manchester Railway, uncertain whether to use locomotive or stationary engine haulage, announced competitive trials. There were four steam-locomotive entrants: *Rocket*, promoted by George and Robert Stephenson; *Novelty*, the work of William Braithwaite and John Ericsson; *Sans Pareil*, by Timothy Hackworth; and Timothy Burstall's *Perseverance*. There was also a horse-on-treadmill entrant known as *Cyclopede*, submitted by John Brandreth – a sceptic with a sense of humour – and a 'man engine' entered by a member of the Winans family.

The 0-4-0 *Locomotion No. 1*, built in the Stephensons' Forth Bank Works, Newcastle upon Tyne, and delivered in 1825 as *Active* to the Stockton & Darlington Railway – the first to introduce a proper passenger-carrying service on 27 September 1825 when an eighteen-seater coach called *Experiment* was introduced.

Timothy Hackworth was born in Wylam on 22 December 1786, the son of John Hackworth (1746–1802/1804), foreman blacksmith in the local colliery. He is chiefly, if unluckily, remembered for his *Sans Pareil* of 1829: defeated comprehensively by Stephenson's *Rocket* in the Rainhill Trials. Succeeding to what had been his father's job in 1807, Hackworth had soon been added to a team assembled by Christopher Blackett, owner of the colliery, which also included the viewer, William Hedley, and Jonathan Forster, a 'wright'. The men were to investigate the possibility of replacing horse traction with steam. Hackworth left Wylam in 1816 after differences with Hedley, but was soon working for Walbottle Colliery. A year spent working for Robert Stephenson & Company persuaded the Stephensons to recommend Hackworth as locomotive superintendent of the Stockton & Darlington Railway, where he was to stay for many years. Hackworth is credited with ensuring that the first locomotives, including *Active* (now known as *Locomotion No. 1*), were fit for purpose and he designed 0-6-0 *Royal George* of 1827 in which an efficient blast pipe was used for the first time. Locomotives were built in collusion with his brother, Thomas, and Nicholas Downing, but the partnership was dissolved in 1840 and Hackworth continued alone. Technological progress soon overtook the basic 0-6-0 design with characteristic rear-mounted cylinders. A more efficient 2-2-2 was built in 1849 to the designs of John Gray, locomotive superintendent of the London & Brighton Railway, but Hackworth died on 7 July 1850 before anything more could be done. Hackworth and his wife, Jane Golightly, had nine children; their son, John Wesley Hackworth (1820–91), continued his father's work after 1850 and is credited with the 1859 patent Hackworth valve gear.

The victory of Stephenson's *Rocket* at Rainhill ended the formative phase of railway locomotive development, and the influence of the beam engine disappeared almost overnight. Drive was instead taken from the cylinders by connecting rods attached directly to crankpins on the wheels.

Rocket, however, was soon improved. The thrust of the original pistons, which were set an angle of about 38 degrees to the rails, tended to lift the wheels clear of the rails at high speed. Consequently, on the engines delivered to the Liverpool & Manchester Railway in 1830, the cylinders were set all but horizontally. *Rocket* was rebuilt to similar standards, acquiring new cylinders inclined at just 8 degrees to the horizontal.

The opening of the Liverpool & Manchester Railway in 1830 was a great public occasion, as the genuinely competitive nature of the Rainhill Trials and the performance of *Rocket* had been a spectacular public relations success. But the day was notable for the first fatality on a public railway when the Member of Parliament for Liverpool, William Huskisson, unwisely engaged in conversation with the Duke of Wellington while standing between the two parallel tracks and was struck down by *Rocket* after falling in his attempt to board the Duke's carriage. The dying Huskisson was duly raced to Eccles in search of aid by *Northumberland*, which is said to have reached 37mph – a world record for steam traction. The railway boom had begun in adversity.

A modern full-size replica of Stephenson's 0-2-2 *Rocket*, the locomotive that did as much as any other single machine to establish the idea of railways in the public consciousness. The illustration shows the locomotive as built, with steeply inclined cylinders.

The use of two cylinders operating out of phase was well established even by 1830, as it allowed the engine to start even if one crank had stopped on dead centre and smoothed the transmission of power to the wheels. Gearing, jackshafts, bell-crank levers and other unusual power transmission systems were all tried in this formative period, usually without lasting results. Richard Roberts had tested a bell-crank on *Experiment*, built for the Liverpool & Manchester Railway in 1833, and then applied it to the 2-2-0 *Britannia*, *Hibernia* and *Manchester* supplied by Sharp, Roberts & Company to the Dublin & Kingstown Railway in 1834. A broadly comparable 4-2-0, *Earl of Airlie*, was made by J. & C. Carmichael for the Dundee & Newtyle Railway in the same period. Bell-cranks allowed a vertically moving piston clearly adapted from table-engine practice to work a horizontal crank, but were unable to demonstrate any superiority over conventionally placed cylinders (though the failure of *Earl of Airlie* has sometimes been ascribed more to the poorly designed piston valves than to the bell-crank itself). The 2-2-0 *Vauxhall*, a conventional crank-driven locomotive made by George Forrester in 1834, soon proved superior to the bell-crank engines on the Dublin & Kingstown.

Timothy Hackworth's *Royal George* of 1827 was the first locomotive to have three axles coupled by a rigid rod.

The first locomotive to be designed with inside cylinders driving on to a crank axle was probably Timothy Hackworth's *Globe*, ordered by the Stockton & Darlington Railway in February 1830; the earliest to be built, however, may have been *Liverpool* by Edward Bury. Yet the first engine of this type to be truly successful was the work of Robert Stephenson. The 2-2-0 *Planet* was ordered by the Liverpool & Manchester Railway in the summer of 1830, run experimentally on 23 November, and then set to haul an 80-ton load from Liverpool to Manchester on 4 December. The journey was completed successfully in two hours and fifty-four minutes. *Planet* was a small engine of about 8 tons, with a multi-tube boiler, and wheel diameters of 3ft 1in (leading) and 5ft (driving). Placing the two 11 × 16in cylinders under the smokebox not only reduced heat losses but also increased efficiency.

The easiest way to improve the performance of a 'Planet'-type engine was to enlarge the boiler, requiring an additional carrying axle beneath the firebox. 0-4-2 derivatives of the basic 2-2-0 pattern appeared in the autumn of 1832, but the perfected design was the subject of English Patent 6484, granted to Robert Stephenson on 7 October 1833. The trailing axle allowed a larger firebox and boiler, with enlarged heating surfaces and better steam-raising capabilities. A steam brake was fitted and flangeless driving wheels were intended to avoid breakage of crank axles.

The first engine, *Patentee*, was eventually acquired by the Liverpool & Manchester Railway in April 1834 for £1,000. The locomotive had two 11 × 18in cylinders and weighed 11 tons 9cwt without its four-wheel tender. The first 0-6-0 of this form, intended for banking and heavy freight, was supplied by Stephenson to the Leicester & Swannington Railway in September 1833; it had 4ft 6in-diameter coupled wheels, two 16 × 20in inside cylinders, and weighed about 17 tons.

The success of even the earliest railways increased the loads engines were expected to pull. This was answered simply by making them bigger; boilers increased in diameter and had more fire tubes, increasing the amount of water that could be turned to steam and allowing cylinders to be enlarged; boiler pressures rose, forcing the development of adequate safety valves; driving wheels were enlarged so that the distance travelled with each rotation grew.

Stability problems were solved by moving cylinders from vertical, then diagonal to horizontal; by adding axles and, particularly, in the development of bogies or pivoting trucks to guide engines around curves. The first bogie had been patented in 1812 by William Chapman; Isaac Dripps of the Camden & Amboy Railroad was the first to fit a pilot (or 'cowcatcher') and a pony truck, when, in 1832, he converted a Stephenson-made 0-4-0 to 2-2-2.

For all these changes, however, the inside-cylinder design pioneered by *Planet* in 1830 remained practically unchanged in Britain into the twentieth century. The most important innovations in this period concerned the valve gear, which allowed the expansive properties of steam to be used to their best effect. This increased power while often reducing the consumption

A replica of the 2-2-2 *Adler*, built in England in 1835 by the Stephensons, which was the first steam locomotive to run in Germany.

of coke. The valve-gear system patented in 1842 by Williams & Howe is now better known as 'Stephenson's', though Stephenson himself always tried to give credit where it was rightly due. It was used until the very end of the steam age alongside a rival mechanism patented in Belgium in 1844 on behalf of Egide Walschaerts.

By 1840, the idea of railways had caught public attention not only in Britain, which can justifiably be seen as the 'cradle of the railway', but also in Europe and North America. Though enterprising engineers such as Marc Seguin in France and John Stevens in the USA had already produced their first designs by the time of the Rainhill Trials, the first public railways in France (opened in July 1830), Germany (December 1835 in Bavaria), Belgium (May 1835) and the Netherlands (1839) were all supplied with British-made or British-inspired motive power. The first Austrian and Russian railways, both dating from 1837, relied on locomotives supplied from the USA – usually by Norris or Harrison, Winans & Eastwick. Not all of the projects were successful but, gradually, the lure of railways spread.

Ross Winans, born on 17 October 1796 in Vernon Township, New Jersey, to William and Mary Winans, is among the most notable personalities of the early American railroad history. An early interest in engineering led him to design a wheel with outside bearings, and this in turn brought him to the attention of the embryonic Baltimore & Ohio Railroad. There he worked with Peter Cooper on the creation of the locomotive *Tom Thumb* and became assistant engineer of machinery in 1831. Among Winans' earliest patents were one for the construction of axles and bearings, and another for a 'double truck car' (which he did not invent, but was merely the first to protect!). Opinionated and querulous, Winans left B&O in 1835 to enter partnership with George Gillingham, lease the B&O workshop in Mount Clare, and make locomotives and machinery. The partnership split in 1841, leaving Winans to build a new factory alongside the Mount Clare workshop. The Baltimore & Ohio and the Philadelphia & Reading Railroads proved to be Winans' biggest customers, purchasing nearly 70 per cent of 267 locomotives built between 1843 and 1863. Winans had dissociated himself from railway work after an argument in 1857 with the superintendent of the B&O, Henry Tyson: the railway demanded leading bogies on its locomotives, something with which Winans disagreed strongly. Winans was among the first to substitute coal for wood as fuel, and developed quirky 'Crabs', 'Mud Diggers', and a variety of 'Camels' with the cab perched midway along the top of the boiler so that an elongated firebox could be incorporated. He even built 4-8-0 *Centipede* (1854). Never one to accept the ideas of others, Winans fitted a valve gear of his own design which opened and closed the ports almost as rapidly as the better-known Corliss. He produced an extraordinary series of semi-submersible 'cigar ships', but his anti-Federal sympathies in the Civil War damaged his reputation severely – even though he had had little to do with the 'Confederate Steam Gun', invented by Charles Dickinson and made in Boston, which had simply been adjusted in the Winans factory on its way to 'defend Baltimore'. Winans, and particularly his son, Thomas DeKay Winans (1820–78), helped to create the imperial railway between Moscow and St Petersburg. Married twice, to Julia de Kay and then Elizabeth West, Winans died in Baltimore on 11 April 1877; he had five children. His estate is said to have amounted to $20 million.

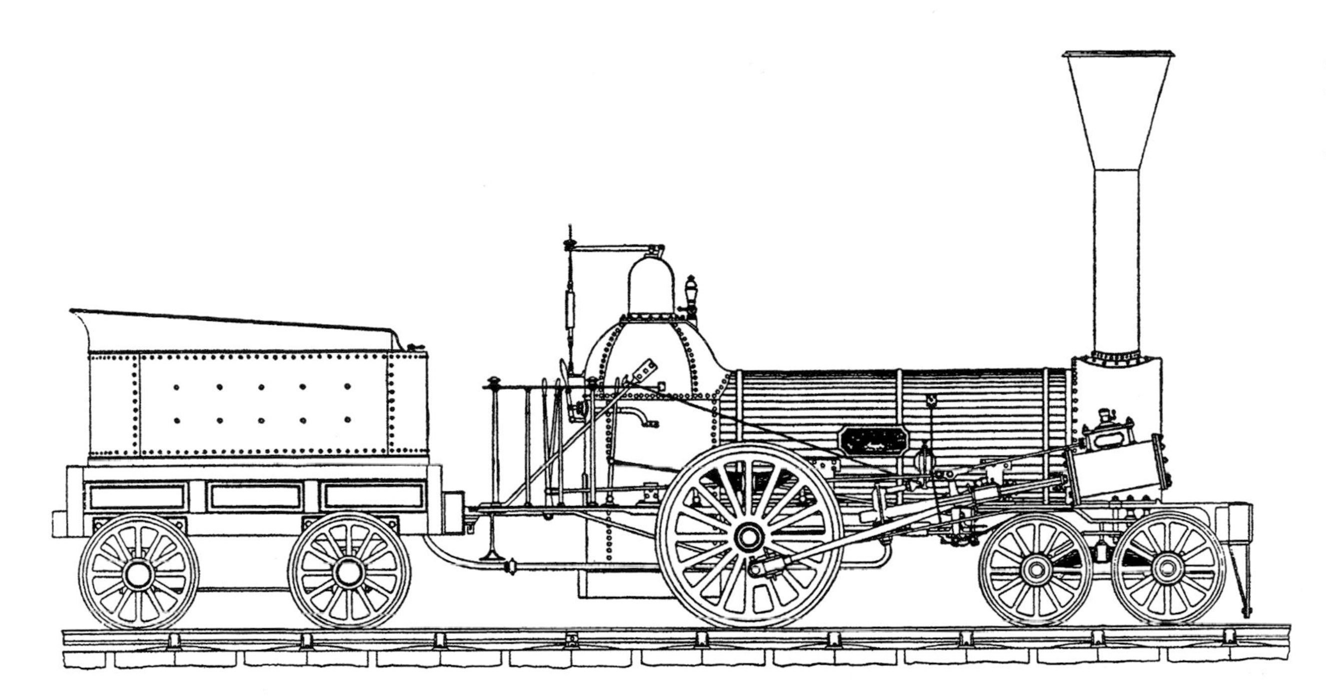

A typical Norris-made 4-2-0 of Class B, made for the Birmingham & Gloucester Railway in 1839. The drawing, possibly inaccurate in small details, comes from P.R. Hodge, *The Steam Engine* (1840).

In Britain, this took the form of 'Railway Mania'. In 1825, the British parliament unwisely repealed the Royal Exchange and London Assurance Corporation Act of 1719. This had been passed in the aftermath of the 'South Sea Bubble', a promotional scheme of dubious morality that had raised huge sums of money, largely from the aristocracy and the mercantile classes, then collapsed to the ruination of many investors. Repeal removed the necessity to obtain an Act of Parliament before a joint-stock company could be formed. This played straight into the hands of the speculators and unbelievable numbers of railway projects were developed. Some had a basis in reality, but at least as many were groundless and often fraudulent speculations. In 1846, Parliament passed 272 railway acts seeking to build about 9,500 miles of track at a time when less than half of this existed across Britain. Only 3,945 miles were in operation by the year-end.

Among the leading lights was George Hudson (1800–71), 'the railway king', who had risen to eminence from a mere proprietorship of a drapery business to ownership of the York Union Banking Company. Hudson was a major shareholder in several railway companies, including the York & North Midland and the Birmingham & Derby Junction Railway, and had been the active promoter of many others. He had been instrumental in the formation of the Railway Clearing House to oversee the distribution of revenue. However, he was also guilty of bribery, buying votes in parliament and false accounting.

His fall was sudden, rapid and terminal; after a short period of imprisonment for debt, he died in penury. Railway Mania collapsed with Hudson, bringing the banking system to the verge of collapse and threatening the very basis of the British economy.

Some of the schemes that collapsed in the mid-1840s would reappear much more successfully at a later date. And though the nascent railway industry was thrown into temporary recession, track mileage was destined to grow greatly in the decade that followed.

2

POWER AND SPEED

Technological Progress and the Growth of Railways, 1846–89

By the time Railway Mania had subsided, and the most risky propositions had been deferred, reinterpreted or abandoned, the design of the locomotive had stabilised. Quirky features such as vertical cylinders and bell-crank drive had been discredited by two decades of practical experience. The coke-burning fire-tube boiler had become standard, cylinders were placed horizontally and stability had been improved from the 0-2-2 of *Rocket* and the 0-4-0 of *Sans Pareil*, in almost all but the 'long boiler' locomotives, by the addition of at least one axle.

Though the Great Western Railway clung to broad gauge (7ft 0¼in) and some railways in Ireland remained wedded to 5ft 3in or 5ft 6in, virtually all British railways – industrial lines excepted – had accepted the 'standard gauge' of 4ft 8½in. There was less consensus in Europe until the idea of countrywide networks became established. Most European countries had adopted standard gauge by the mid-1850s, but Ireland and Spain clung to 5ft 6in and Russia, after starting with 6ft, settled on 5ft. Even North America, where the profusion of railroads and their geographic separation was unmatched elsewhere in the nineteenth century, generally accepted the standard.

Advances in technology in the 1850s and 1860s were generally confined to details, such as the introduction of split rings on solid-head pistons, credited to John Ramsbottom. Henri Giffard developed the first practicable injector, and the prolific Ramsbottom then produced his duplex-spring safety valve. Trials were also undertaken successfully to enable cheap coal to be used instead of expensive coke, the change being almost entirely due to improvements in firebox design.

Most newly built British locomotives of the early 1850s had three axles: 2-2-2, with large driving wheels and a carrying axle at each end, and the 0-6-0 with all three axles coupled. The 2-2-2 configuration was intended for passenger trains, which had been demanding ever greater speed once experience had discredited the claim that people's heads would explode at greater than 40mph. The perfected 0-6-0, with its origins in Hackworth's *Royal George* of 1827, was used largely as what was often called a 'merchandising engine' (but is now generally known as a goods engine) with six small-diameter wheels coupled together.

The differences between the two classes were easily explained. The passenger trains of the 1850s, excepting the broad gauge of the Great Western Railway, were composed largely of small four-wheeled carriages amounting to no more than a few tens of tons behind the tender. These could be moved at considerable speed, as the 'Battle of the Gauges' proved. The broad-gauge engines proved to be easily capable of 60mph,

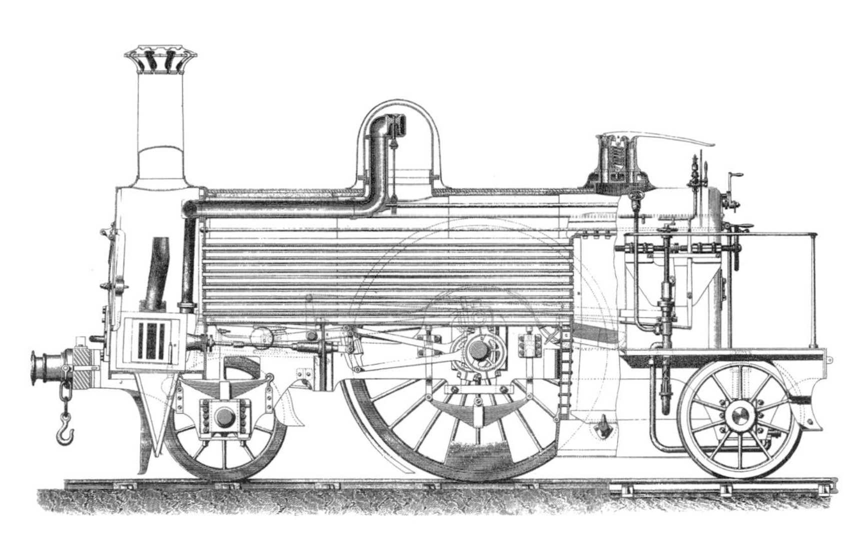

An LNWR 2-2-2 Express Passenger Engine by John Ramsbottom. (Official *Record of International Exhibition, 1862*)

The London & North Western Railway 2-2-2 *Cornwall* is seen here towards the end of a career that spanned several decades. The machine was originally built in 1847 with a squat small-diameter boiler that allowed the driving axle virtually to pass over it. This design was intended to improve stability, but proved to be a dismal failure. The engine was rebuilt in 1858 and acquired a cab at a later date.

and even on the standard gauge, where maintaining stability was more problematical, speeds of 45mph were by no means uncommon. Goods trains, though much slower, were very much heavier.

Pulling a light train quickly was easily achieved by locomotives with a single driving axle, allowing wheels of large diameter to be used and keeping friction to a minimum by eliminating coupling rods. That only a proportion of the weight of the locomotive was applied to the driving wheels did not become a problem until train weights increased dramatically in the 1880s and 1890s. Indeed, the heyday of the 'Single-Wheeler', as 2-2-2 and 4-2-2 express locomotives were often called, did not pass until the very end of the century.

The goods train needed a locomotive that could exert far more tractive effort, and this was best done by allowing the entire weight of the locomotive to rest on the driving wheels, which were coupled together to ensure they rotated in unison, and to keep the wheels as small as practicable. Small-wheeled locomotives were obviously slower than those with large wheels, simply because, for each piston stroke, larger wheels moved a greater distance along the track. The difference was directly proportional to wheel diameter. A goods engine with a 4ft 6in wheel would move about 14ft 2in during a single rotation, whereas an engine with an 8ft wheel would move 25ft 2in. All other factors being equal, a passenger locomotive with 8ft wheels would attain about 36mph when a goods engine with 4ft 6in wheels was running at only 20mph. In both machines, pistons and associated valve gear would be moving at the same rate.

This encouraged the development and introduction of some of the most extraordinarily large wheel designs. As early as the 1830s, Thomas Harrison had produced a freak locomotive, *Hurricane*, which had driving wheels with a diameter of no less than 10ft. Unfortunately, the wheels and cylinders were carried on a leading carriage with the boiler and the driver on another. Most of the weight that should have contributed to adhesion was concentrated on the unpowered boiler-truck and *Hurricane*, in the words of one writer, 'failed to move as much as itself'.

Fitting large wheels is not difficult, but constraints imposed by the meagre dimensions of platforms, signals, overbridges and tunnels (the 'loading gauge'), especially in Britain, proved to be impossible to overcome. Of course, there were always inventors prepared to try. Isambard Kingdom Brunel proposed a 2-2-2 with a 10ft driving wheel and two 14 × 20in cylinders; *Ajax* and *Mars* being built by Mather, Sons & Platt in 1837. The disc-like driving wheels were made of iron segments, but the locomotives performed so poorly that they were rebuilt immediately with conventional 8ft drivers. *Ajax* lasted only until 1840, running fewer than 16,000 miles at withdrawal.

Francis Trevithick, son of Richard, introduced the 2-2-2 *Cornwall* to the LNWR in 1847; eight years later, the Frenchmen Blavier and Larpent promoted *L'Aigle*, 2-4-0 No. 261 of the Compagnie de Chemins de l'Ouest. Trevithick's locomotive had wheels of 8ft 6in, while those of *L'Aigle* had a diameter of 2.85m (9ft 4½in). Trials showed *Cornwall* to be unsteady, and an additional carrying axle made the locomotive a 4-2-2: the guise in which it was shown at London's Great Exhibition of 1851.

Cornwall and *L'Aigle* both worked after a fashion, but each faced a major constructional problem. No locomotive can work effectually if the boiler cannot supply enough steam to the cylinders continuously. If the boiler is too small (or the cylinders are too large), the system is incapable of developing enough power when demands rise beyond a certain point: something that affected locomotives produced in the first few years of the twentieth century, when the experience of others should have been a lesson to their engineers.

Enlarging the wheels commensurately raised the driven axle above rail height. As the comparative narrowness of standard-gauge track tended to compromise stability if the centre of gravity of rolling stock was raised too far, boilers could no longer be mounted above the axles. Trevithick fitted a small-diameter boiler that was recessed to allow the driving wheel axle to pass over it; the Frenchmen fitted a boiler with limbs projecting forward above and below the driving wheel axles. Neither method proved to be satisfactory, as the steaming rates were too poor to allow prolonged operation in adverse conditions – into the wind, or up a gradient. *L'Aigle,* never accepted for service, was scrapped rapidly. The parsimonious LNWR rebuilt *Cornwall* as a conventionally boilered 2-2-2 in 1858, and the engine, which proved to be fast and steady, continued to run reliably for many years.

Outlandishly large wheels placed beneath the boiler proved to be impossible to justify, as there were better methods of attaining speed. Yet the Bristol & Exeter Railway purchased 4-2-4 tank engines from Rothwell & Company with 8ft 10in drivers. One of these is said to have attained 82.8mph, running downhill, but the large wheels, placed centrally, were adequately accommodated within the broad gauge. Another method, better suited to the standard gauge, was offered by Thomas Russell Crampton (1816–88), to whom a relevant English patent, No. 11349, was granted on 25 August 1841. These locomotives – usually of 4-2-0 type – were intended specifically to pull comparatively modest loads at high speed. The driving axle lay immediately behind the firebox, at the rear of the frame, where it could be turned (in the earliest designs at least) by outside cylinders placed between widely spaced carrying axles.

The first of these engines was made for the British-owned Liége & Namur Railway (in Belgium) by Tulk & Ley of Whitehaven, Cumbria. Completed in 1846, this 4-2-0 had 7ft-diameter driving wheels and cylinders measuring 16 × 20in. Crampton-type engines were then tested on the London & North Western Railway, culminating in the unique 6-2-0 *Liverpool*, built in 1848 by Bury, Curtis & Kennedy. This 35-ton monster, with 8ft drivers and two 18 × 24in cylinders, allegedly attained 75mph with a light train. Unfortunately, the rigidity of its wheelbase was said to have damaged the light and poorly laid track, and the reputation of the Crampton system in Britain suffered as train loads increased.

No one doubted that the Cramptons were fast, but the factor of adhesion inherent in a single axle placed so far to the rear of the boiler-mass was too low to pull anything but a small load. Their greatest

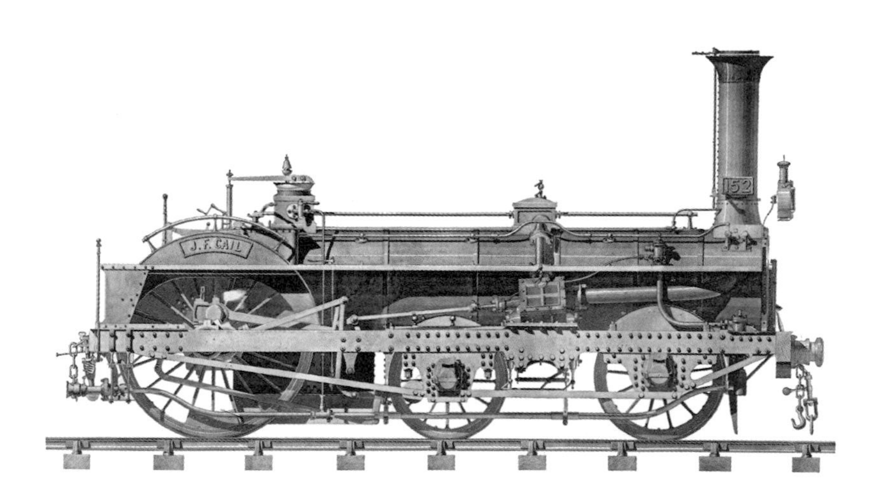

Crampton locomotives were characterised by outside cylinders, a long wheelbase and a single driving axle behind the firebox. They were exceptionally fast for their day, setting many speed records, but prone to ride harshly. They were also poor hauliers and were customarily restricted to light high-speed passenger trains. This French-made 4-2-0 example was the work of Cail of Lille.

success proved to be in France, where the phrase *Prenez le Crampton* ('take the Crampton') was synonymous with 'take the train' for many years. This was partly due to differing traffic conditions. In England and Wales, people had tended to move to urban districts as commerce grew from the effects of industrialisation; 1850s France, despite long-established academic and scientific traditions, was still largely rural. Train loads in Britain grew rapidly to satisfy the burgeoning 'commuting classes', and the short distances between many of the stations favoured locomotives that could start-away reliably.

French stations could be few and far between, and owing to the size of the country (about four times the area of England and Wales), trains were lightly loaded and could make the best of long straight sections of track. Cramptons, for a time at least, were ideally suited to this type of service. They were made in large numbers by companies such as Cail, were modernised when necessary, and survived on branch lines almost until the twentieth century. It has been estimated that 230–250 Cramptons were made in Europe, split almost equally between France and Germany.

Ironically, when unofficial speed trials were undertaken by the British LB&SCR 0-4-2 *Gladstone* (82mph) and the SER 4-4-0 *Onward* (81mph running as a 4-2-2) on the Chemins de Fer du Nord at the end of the Paris Exposition of 1889, an old French locomotive beat them both – attaining the highest authenticated speed attained by a French Crampton, 144kph (89.5mph) with a train of more than 150 tons, which would have been a world record for steam traction at that time.

A last throw of the large-wheel die could be seen at the 1889 Paris Exposition, when *La Parisienne* of 1886 was shown. Designed by Auguste Estrade (1817–92), a rich *propriétaire viticole* from Languedoc with an engineer's training, the 0-6-0 was made by Boullet & Cie of Paris. It had two outside cylinders measuring 47 × 70cm (18½ × 27½in) and driving wheels with a diameter of 2.5m (8ft 2½in), matched by the tender wheels and even by the wheels of the special double-deck passenger carriage. Inspired by the Crampton, Estrade had been working on his ideas since the 1850s and is believed to have sought a patent in 1884. The locomotive was commissioned in the vain hope that it would achieve 120kph, but, after an uninspiring test run or two, *La Parisienne* was consigned to history.

The locomotives shown in London in 1862, considered as a group, contained several improvements over those displayed eleven years earlier. A major advance had been the advent of efficient means of burning cheap coal instead of expensive coke. Even though the development was still imperfect, the 1862 engines incorporated 'smoke-consuming' fireboxes, double or divided; air-assisted combustion; brick deflector arches; and a host of other innovative ideas. There were better means of following the curves in track – articulation and radial axles, for example – and the perfection of the Giffard injector, patented in France in 1858, made manually operated feedwater pumps obsolescent. Another major change lay in the rejection of ultra-low centres of gravity, which in 1851 had been considered essential if high speeds were to be attained on standard-gauge track.

Joseph Hamilton Beattie was born in Londonderry on 12 May 1808, the son of an architect. He came to England in 1835 as assistant to Joseph Locke, engineer of the Grand Junction Railway and then the London & Southampton Railway. Beattie became the L&SR carriage and wagon superintendent, but, in 1850, superseded John Viret Gooch (brother of Daniel Gooch) as locomotive superintendent of the London & South Western Railway. There Beattie remained until his unexpected death on 18 October 1871 during a diphtheria epidemic. He is best known as an early exponent of feedwater heating (1854), and for a lengthy series of experiments – and many patents – concerned with the development of fireboxes and combustion chambers that would burn inexpensive coal instead of more specialised coke. Beattie, unlike most of his contemporaries, could see that the future lay in coupled wheels and commissioned no single-wheelers after 1859; the Great Western, the Midland Railway and others were still building them in the 1890s. Joseph Beattie, married to Ellinor Morrison, had six children; he was succeeded by his son, William George Beattie (1841–99), locomotive superintendent of the LSWR in 1871–78.

The catalogues of the 1862 exhibition reveal exhibits ranging from the cheapest contractors' engines to oddities emanating from France. Some observers detected fundamental differences between engineering practice in Britain (where speed was often paramount but adhesive weight was not) and central Europe, where slow speed slogging up lengthy banks required small wheels and as much adhesive weight as possible. Consequently, the major British exhibits included a 2-2-2 express passenger engine designed by John Ramsbottom for the London & North Western Railway. Built in the company workshops in Crewe, the locomotive had driving wheels with a diameter of 7ft 7½in, a fixed wheelbase measuring 15ft and two 16 × 24in cylinders mounted externally.

The Civil War raging in the USA prevented the inclusion of transatlantic exhibits, but the European manufacturers arrived in force. Among them were some remarkable achievements, though British journalists tended towards disparaging comments on the grounds that the 'foreign' designs were either too complicated or failed to follow the established chimney first/cab last layout.

The Austrian section included *Duplex*, a 4-2-0 designed by John Haswell for the Royal Austrian State Railway. Its characteristics included 6ft 8½in driving wheels and a short fixed wheelbase, owing to the absence of a bogie, of about 11ft. Four cylinders provided the locomotive's most unusual feature. Mounted externally, each had an 11in bore and a 25in stroke; the engine weighed about 32 tons in working order, 12.5 being carried on the driven axle.

Haswell placed the cylinders outside the frames, each pair driving on to a double-ended crank attached to driving wheels that were carried inside the plate frame. The goal was to eliminate vibration by balancing the rotational forces created in a normal two-cylinder design. To demonstrate his ideas to sceptical observers, *Duplex* was tested by running at 400rpm while suspended above the ground – the equivalent of 96mph! The horizontal oscillation was merely one-twelfth of an inch, vertical movement being one-fifth. The engine had been tested on the Austrian railway at 70mph, proving, as predicted, to run with praiseworthy steadiness, but it was too complicated to be durable and had a short service life.

John Haswell was born in Lancefield, near Glasgow, on 20 March 1812. Apprenticed to a mechanical engineer, he then studied at Anderson's University and worked for some years for William Fairbairn & Co. In 1837, he prepared plans for the repair shop of the Wien-Raaber railway, which was built in 1839–40 under his supervision. Haswell then became manager of the factory and was given responsibility for new construction. His many innovations included *Fahrafeld* of 1846, the first six-coupled locomotive to run in Austria; *Vindobona* of 1851, an entrant in the Semmering competition and inspiration for the Engerth locomotives; *Wien-Raab* of 1855, the first eight-coupled freight locomotive; *Steierdorf* of 1861, the first to be fitted with a steam brake; and *Duplex* of 1861, the world's first four-cylinder design. Haswell also developed a corrugated firebox and made many improvements in production machinery. He resigned in 1882 and died in Vienna on 8 June 1897. Now all but unknown in Britain, the esteem in which he was held in Austrian railway circles is mirrored in the monument above his grave. He married Sophia Lane (1821–1910) and had two children, Alexander Elliot and Mary Fanny Haswell. Neither child seems to have left descendants.

Much stranger than *Duplex* was a group of French locomotives credited to Jules-Alexandre Petiet (1813–71), locomotive superintendent of the Compagnie de Chemins de Fer du Nord in 1845–65 and then principal of the engineering college of l'École Centrale de Paris. Made by Ernest Gouin & Cie of Paris, there were two freight engines (one with four axles, another with six) and a five-axle express-passenger pattern. The freight engines were effectively 0-4-4-0 and 0-6-6-0, with their axles in two separately coupled groups, but the passenger machine was an 0-2-6-2-0 with inside frames and

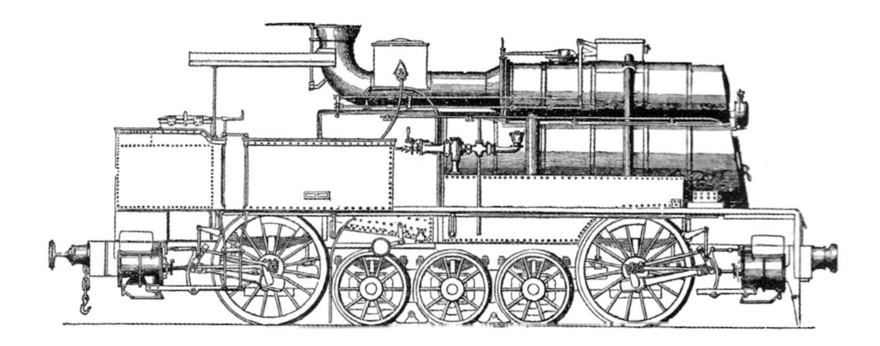

The anachronistic 0-2-6-2-0 passenger locomotive designed for the Nord railway of France by Jules Petiet, exhibited in London in 1862 to much adverse (but somewhat misplaced) comment. (Official *Record of International Exhibition, 1862*, courtesy of the Science Museum, Crown Copyright)

a single driven axle, at each end, separated by a three-axle 'carrying group' mounted rigidly in the frame. The diameter of the driving wheels was about 1.6m (5ft 3in), and each driven axle had its own pair of outside cylinders measuring 36 × 34cm (14¼ × 13½in). Boiler pressure was 8.1at (120psi) and the engine weighed 48 tonnes (47.5 tons) in working order.

Among its oddest features was the chimney, which was carried backward along the top of the boiler to turn upward above the cab. In addition, the boiler, clad in wooden insulation, was protected by highly polished brass sheets. The double-ended nature of construction enabled the locomotives to run in either direction with equal facility, and they also featured an effectual steam drier, a precursor of the superheater.

The Petiet locomotives attracted scorn from the British engineering journalists. Service revealed fundamental flaws, particularly in the 0-2-6-2-0 where poor distribution of weight compromised tractive effort and allowed the bearings of the three central carrying axles to overheat. The eight passenger locomotives proved to ride unsteadily at speed and were rapidly relegated to branch-line services; they had all been scrapped by the mid-1870s. Though twenty 0-6-6-0 Petiet locomotives were made in 1862–64, they were converted to forty 0-6-0 tank engines in the La Chapelle workshops ten years later. It is sometimes claimed that a few Petiets lasted long enough in service to be featured in postcards published early in the twentieth century, but the photographs of the double-enders were at least 30 years old. Close examination of other images reveals conventional Petiet 0-8-0 freight engines reduced to the status of shunters.

The absence of any US presence in 1862 was at least partly due to a perceived British bias towards the Confederacy – many of the blockade runners had come from British shipyards – and obscured the huge strides that had been made in North America. Though a handful of locomotives had been imported from Britain in the earliest days, often unsuccessfully, American engineers had been very quick to follow the lead. A formative period in which designs such as *De Witt Clinton* and *Tom Thumb* had been introduced soon gave way to a creative explosion.

The distances to be covered as the Union expanded westward placed a premium on endurance and also, as the importance of railways increased, on haulage capacity. One result was that the locomotive grew far quicker than its European equivalent. While the British railways were happily building 2-2-2 passenger and 0-6-0 goods engines in 1862, the 'Ten-Wheeler' (4-6-0) and even the 'Consolidation' (2-8-0) had taken to the rails in the USA. Track mileage had increased from less than thirty in 1839 to about 31,000 in 1861.

North American locomotives customarily had two outside cylinders driving on to the leading coupled axle, which gave a characteristic gap between the bogie and the drivers. This allowed a boiler of surprising length to be fitted and the engines steamed well. The frames were usually of bar construction, not the plates favoured in Britain; cabs offered the crews a fair amount of protection in conditions that could range from arid plains to snowbound mountains; large smoke arresters were often prominent, to prevent prairie fires; cowcatchers or pilots guarded against obstructions on tracks (natural or otherwise!); and livery was often a riot of colour. It was by no means unknown for enamelled or painted plaques to record the image of the founder or financier of a railroad, and ornamental metalwork could fill the space between the driving wheels.

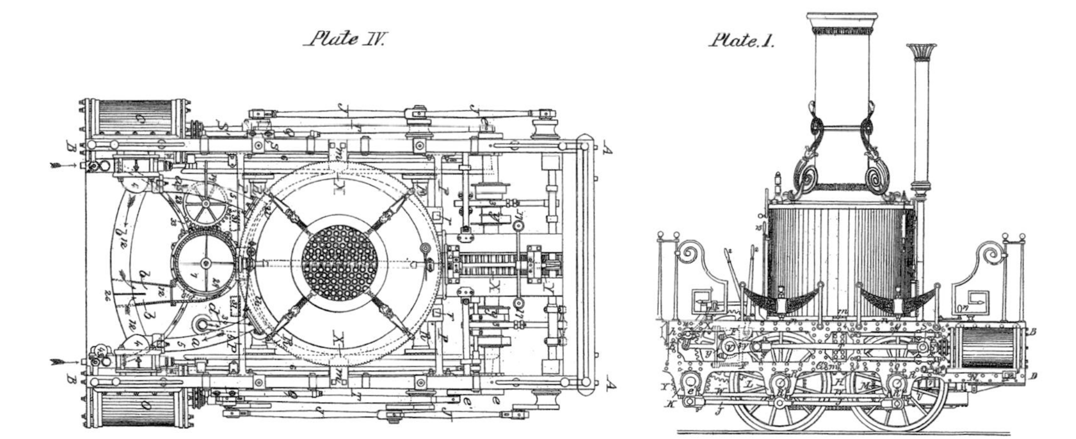

This compact vertical-boilered 0-4-0 was the subject of US Patent No. 308, granted on 29 July 1837 to Ross Winans. (US Government Patent Office, Washington DC)

A typical 4-4-0 American, *W.H. Whiton*, was made by William Mason for the US Military Railroad during the Civil War. Whiton was assistant to the chief of the Military Telegraph Department. The photograph was taken by A.J. Russell on 20 April 1865. (US Library of Congress)

William Mason, not to be confused with the gun designer of the same name, was born in Mystic, Connecticut, on 2 September 1808. The son of blacksmith Amos Mason, he was largely self-trained in a local textile mill. However, Mason was a first-class mechanic and had soon introduced improvements. He is generally credited with installing the first powered loom to operate in North America and patented a 'self-acting [spinning] mule' in 1840 and 1846. He also made railway locomotives from 1853 onward, which included long-wheelbase 4-4-0s with horizontal cylinders and attention paid to their appearance; Mason products are generally regarded as the most elegant of those produced in the USA in the mid-nineteenth century and were among the first to have lead-balancing weights cast into the hollow rims of driving wheels. Mason introduced Walschaerts' valve gear to the USA in 1876 (it never became popular) and made a number of 'Mason Fairlies' prior to his death in Taunton, Massachusetts, on 21 May 1883. Married to Harriet Augusta Metcalf (1826–80), William Mason had several children.

By 1862, the nuclei of the principal railway companies operating in Britain in 1900 could already be seen, with the possible exception of the Great Central. The 'small-line era' had already given way to the first associations, in which virtually all of the small independent railway companies in each of the major geographic regions were subsumed into one corporate body. For example, the Great Western Railway had been incorporated in 1835, the Midland Railway in 1844, the London, Brighton & South Coast Railway in 1846, and the Lancashire & Yorkshire and Caledonian railways in 1847.

In North America in 1862, there were huge numbers of 'Short Lines' as the move towards aggregation had only just begun; there was considerable territorial opposition to out-of-state initiatives even before the Civil War had begun, and, in addition, the distances involved in inter-state travel were often considered to be excessive by the standards of the day. This was to change with the building of the Transcontinental Railroad.

The idea was not new, as it had been mooted almost as soon as the first tracks had been laid in the USA. The so-called Whitney Bill had been defeated in Congress in 1848, but the creation of Oregon Territory (1849) and the incorporation of California into the Union in 1851 persuaded promoters to try again. The discovery of gold and other minerals in California was a great inducement to those who saw a fortune to be made if only the material could be transported eastward.

The Central Pacific Railroad of California was duly incorporated on 28 June 1861 and, greatly concerned about the effect the lack of communication with the pre-Union western states was having on the Civil War, President Abraham Lincoln authorised the construction of a railroad and telegraph line from the Missouri River to the Pacific Ocean in 1862. The work was entrusted to the Central Pacific, building eastward from Sacramento, and to the newly incorporated Union Pacific Railroad building westward from Omaha.

Work began in 1863. With the help of the so-called 'Hell on Wheels Towns', which – with their saloons and brothels, gamblers and whores moved on railroad flatcars from one site to the next – the railroads inched slowly together. The Golden Spike was driven by Leland Stanford at Promontory Summit, Utah Territory, on 10 May 1869, when Union Pacific 4-4-0 No. 119 and Central Pacific 4-4-0 No. 60 *Jupiter* finally met pilot to pilot.

Lauded as a stupendous feat, which it was, the Transcontinental Railroad was soon embroiled in scandal. Vast sums of money had been made out of scarcely legal land transactions, bribes were commonplace, and allowing the railways to form supposedly independent companies to build, cost and charge for track allowed huge profits to be made by substituting poor material for good or simply by multiplying the supposed costs. The Central Pacific had even persuaded the Californian state assembly to grade the gentle foothills of the Sierra Nevada as mountainous terrain, which naturally allowed premium rates to be charged. Much of the track was poorly laid, and derailments, slipping embankments and collapsing trestles were part of everyday life.

The scandal was at its height in 1873, when the so-called 'Credit Mobilier' fraud was exposed, implicating not only the promoters of the Union Pacific and some of those of the Central Pacific, but also a variety of congressmen and government officials. Yet the worst effects of the scandal had passed when the Centennial Exhibition opened in Philadelphia in 1876, and the value of the Transcontinental Railroad had become clear to all.

In 1872, track mileage in the USA had exceeded 67,000. The Southern Pacific joined with the Texas & Pacific in 1882, then with the Atchison, Topeka & Santa Fe Railroad to complete the route from Kansas City to Los Angeles. In 1883, the Northern Pacific joined St Paul, Missouri, with Portland, Oregon, to provide an alternative route across the continent; and the Canadian Pacific Railway finally opened in 1886.

There was much to admire about the railways in the USA. They had succeeded in bridging the gap between the prosperous east and the sparsely populated but mineral-rich west: blasting a path through the Sierra Nevada, bridging canyons and rivers, subduing the Great Plains. Success could be seen in a railway-engineering industry pioneered by Matthias Baldwin and others who saw the virtues of standardisation instead of drawbacks claimed by (mostly European) detractors, and in the ability to manufacture locomotives, rolling stock and all associated items in large numbers.

Among the locomotives shown in Philadelphia were a 2-8-0 of the Lehigh Valley Railroad, made by Burnham, Parry, Williams & Co. of Philadelphia (the Baldwin Works), and a 4-6-0 'Camelback' from the Philadelphia & Reading Railroad workshops. The Consolidation had 4ft 2in

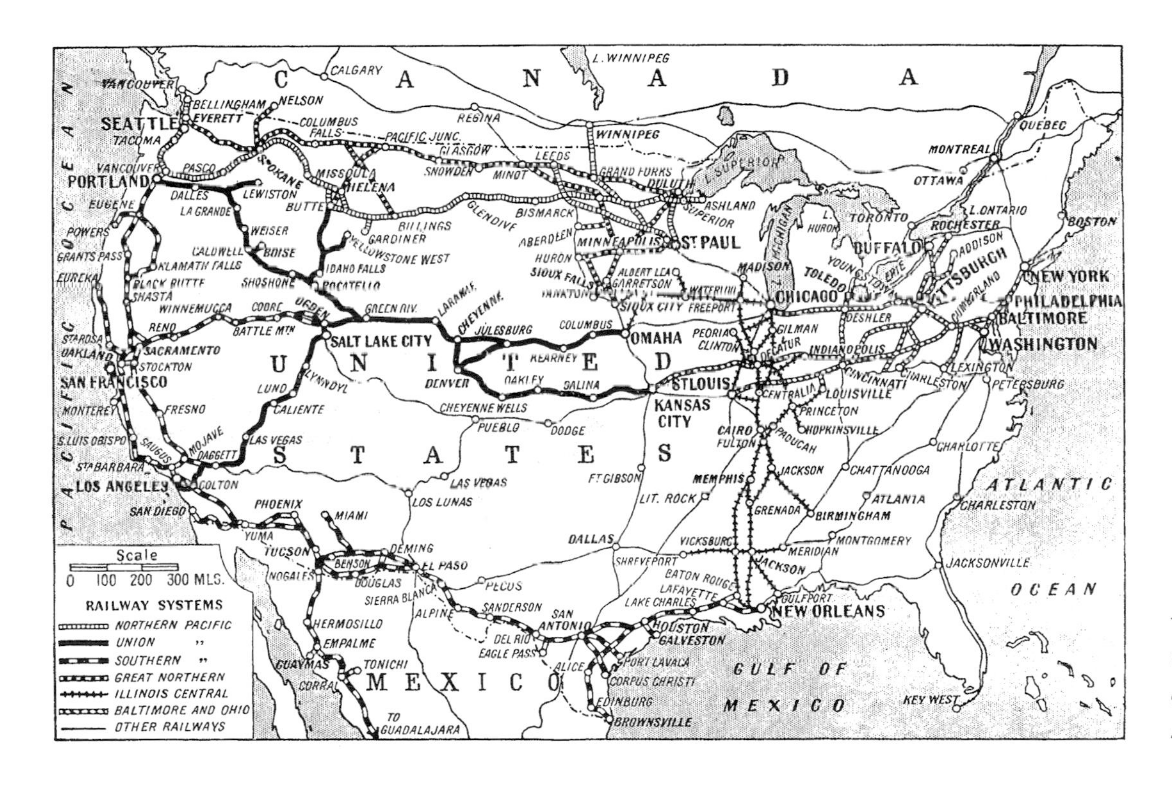

The Golden Spike was driven at Promontory Summit, Utah Territory, in 1869. By the time this map was drawn in the 1930s, railroads criss-crossed North America.

driving wheels, two 20 × 24in cylinders, and weighed 50 short tons without its tender; the Ten-Wheeler had 4ft 6in wheels, two 18 × 24in cylinders, and weighed about 39 short tons without tender. The engines were surprisingly light for the number of axles, but were destined to run on lightly laid track. In addition, their boiler pressures were low: 130 and only 110psi respectively, giving a tractive effort at 80 per cent efficiency of 25,022lb and 12,672lb.

The ease with which even the most outlandish projects could be promoted ensured that the USA was not to be spared embarrassments. In 1881, the Grant Locomotive Works of Paterson, New Jersey, built two curious-looking machines for the Fontaine Engine Company, delivering No. 1 in March and No. 2 in June. The quirky 'wheel-on-wheel' drive had been protected by US Patent No. 230472, granted to Eugene Fontaine of Detroit on 27 July 1880.

Eugene Fontaine was born in Canada in 1835, the son of a French immigrant, and is said to have trained as a railway engineer. He moved to the USA shortly after the Civil War, where he sought a patent for a Cattle Car (1867). Other patents followed: machines to make wire nails and needles, in addition to designs for railroad crossings and railroad signals. Toledo city directories for the 1880s list the Fontaine Crossing & Signal Company, but the business is believed to have lasted only until about 1892. Fontaine, 'inventor' as the 1900 US Federal census had called him, died in Ohio in September 1909.

Fontaine's 1880 locomotive patent sought to lower the centre of gravity and allow faster running without increasing the speed of the piston. The locomotives had two 17 × 24in cylinders; and, according to the *Scientific American* of 5 November 1881, the diameters of the drive-, transfer- and rail-wheels were 72in, 56in and 70in respectively. This gave a 'geared up' ratio of 1:1.28. The notation is either 4–2–2 or 4–4–2, depending on how rigidly the definition 'driving wheel' is implemented!

The Fontaine engines ran trials on several railroads in the northern USA and southern Canada, but their supposed advantages proved to be illusory. Converted to 4–4–0 layout in 1884, they ended their days on the Wheeling & Lake Erie Railroad.

If the Fontaine had a basis in reality, the Holman Truck has often been seen as a deliberate fraud and many writers have simply taken at face value the disparaging comments made in 1907 by Angus Sinclair in *Development of the Locomotive Engine*. Sinclair was a well-trained railway engineer and his low opinion of the speeding-truck scheme was entirely justified; but whether his allegations that the entire concept was a

The 4-8-0 *Champion* of the Lehigh Valley Railroad was built in the company's workshops in Weatherly, Pennsylvania, in 1882.

The Grant Locomotive Works of Paterson, New Jersey, built two of these curious-looking machines for the Fontaine Engine Company in 1881. (From an engraving published in the British periodical *Engineering*)

deliberate sham is open to question. It pays to remember that the Holman locomotive originated at a time when public attention was attracted to a 'railway' to carry 10,000-ton ships across the Panama isthmus, giant steam-powered flying machines, and even the Dynamite Gun.

> **William Jennings Holman**, born in 1819, is listed in the 1880 US Federal census as 'civil engineer', who had spent part of his career building railways. According to the obituary published in the *Scientific American*, he had constructed the Peru & Indianapolis Railroad in the 1850s and had served as its president for many years. Among his children were sons Harold (1868–1920) and William Jennings Holman Jr (1872–1941), both of whom were involved in the Holman Locomotive Speeding Truck Company. Holman died in 1904.

The idea of a 'speeding truck' can be seen as a logical progression from the Fontaine patent, as it allowed virtually any locomotive to be adapted to increase rail speed without increasing piston rate commensurately. Holman purchased an 1887-vintage 4-4-0 from the Chicago, Milwaukee, St Paul & Pacific Railroad (the 'Soo line') in 1894, and trials were successful enough for an application for what ultimately became US Patent 546153 to be made on 18 December 1894; the patent drawings clearly show this particular locomotive. The principal novelty lay in the 'speeding truck', a three-wheel shoe supporting a two-wheel carrier, which in turn supported a comparatively small driving wheel. The assembly had the look of a pyramid of wheels.

The locomotive, which was basically an 'American' (4-4-0), had to be raised much higher than normal to accommodate the auxiliary wheels. Among the goals were an increase in train speed without changing piston speed, a reduction in axle loading, and improvements in economy. Holman suggested that a locomotive fitted with speeding trucks would travel further than a conventionally wheeled example in the proportion 4:7.

Holman and his supporters were confident enough to order a new locomotive from Baldwin, probably in 1896 when the 'Holman Locomotive Speeding Truck Company of Iowa' was duly incorporated in Sioux City. Advertisements placed in *The Philadelphia Times* in February 1897 offered investors the chance to buy $10 million worth of shares in the company: surviving share certificates illustrate the specially adapted Baldwin 4-4-0 pulling a train at speed. Engine No. 10 was duly run on the South Jersey Railroad, beginning in the summer of 1897, and a new patent (US 597557 sought on 7 May 1897) was duly granted to William Jennings Holman, 'Assignor to The Holman Locomotive Speeding Truck Company'.

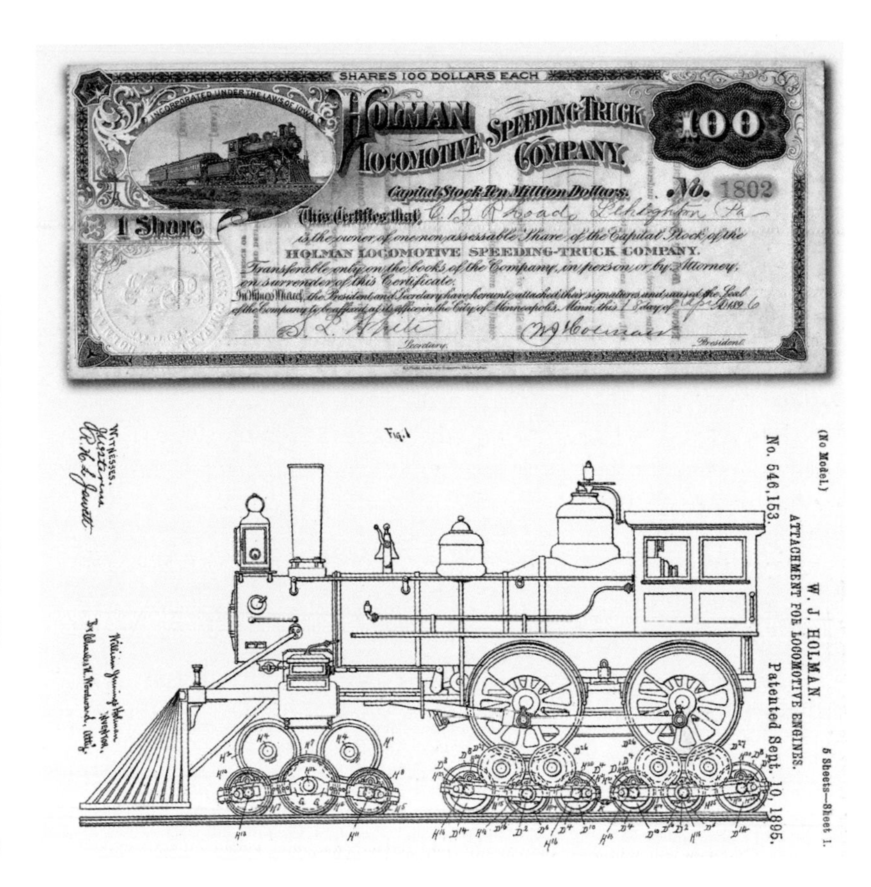

US Patent 546143, granted in 1895 to William Jennings Holman, and one of the share certificates issued to raise funds for the Holman 'Speeding Truck'. (Drawings courtesy of the US Government Patent Office, Washington DC)

Trials predictably showed the speeding truck idea to be flawed. Reports of speeds in excess of 100mph can be discounted, but those that drew attention to the unsteadiness of the assembly were undoubtedly correct.

It was said that the crews feared for their lives, and were ready to jump at a moment's notice. The locomotive was derailed at least once when the trucks, to which it was attached only by gravity, slipped from under a driving wheel. In 1898, No. 10 reverted to standard 4-4-0 form, which Baldwin accomplished merely by changing the bogie and driving wheels, and was sold to the Kansas City & Northern Connecting Railroad.

Though the speeding trucks predictably failed to convince anyone of their merits, it is easy to believe that the promoters had complete faith in them. William Jennings Holman died shortly after the entire project collapsed, but there is no obvious evidence to show that the legitimacy of the shares (unwise though purchase would have been) was ever challenged successfully in court.

The Search for Efficiency

By the middle of the nineteenth century, the advantages of 'compounding' – using steam exhausted from one cylinder to expand in another before being released to the atmosphere – were well established for land engines. Indeed, William McNaught, son of the supposed inventor of the portable engine-indicator, was making good business with what became known colloquially as 'McNaughting': adding new high-pressure cylinders to existing engines. This not only extended their useful life, but also increased power and improved fuel economy.

The idea of compounding had occurred in the 1780s to Jonathan Hornblower, but his engine of 1784 had fallen foul of the 'catch-all' English patent obtained some years earlier by James Watt. This was to be used on many occasions to protect Boulton & Watt's monopoly, stifling many promising developments until Watt died in 1819. However, Cornishman Arthur Woolf had been able to patent a compound engine of his own design in 1805 and substantial numbers had been installed at sites ranging from breweries to tin mines.

Woolf had a shaky grasp of thermodynamics, like virtually all his contemporaries, and the proportions of his high- and low-pressure cylinders were usually ill matched. Yet Woolf engines worked well enough to demonstrate that reduced temperature loss in the cylinders improved economy, and less strain was placed on moving parts. These advantages were to prove attractive to locomotive engineers. The emergence of the Cornish Engine, largely the work of Richard Trevithick, then hindered the development of Woolf engines; but, ironically, they were so successful abroad that virtually any beam engine working in France was known for a time as a 'Machine de Woolf'.

As early as 1845, a proposal was made in England that railway locomotives should have four cylinders, two high and two low pressure, with one of each grouped together so that they could drive a single crankpin. This was the forerunner of 'superposed' designs by Mallet and Vauclain, but there is no evidence to show that it was tried on a railway. Similarly, a British patent granted in 1859 to Salmon of Glasgow to protect a variety of compound systems does not seem to have attracted much attention. Yet it included a proposal for high- and low-pressure cylinders to be grouped at each end of the frame, driving separate crankpins with connecting rods that crossed; one in which the cylinders, superposed centrally, drove the front axle (high pressure) and the rear axle (low pressure) separately; a system in which four cylinders, in line laterally, drove on to a single axle; and a fourth in which a single high-pressure cylinder exhausted into a low-pressure cylinder alongside it within the frames. The third and fourth of Salmon's ideas were to be exceptionally successful when exploited by engineers such as Mallet, de Glehn, Henry and du Bousquet twenty years later.

In 1850, English Patent 13029 was granted to James Samuel, engineer of the Eastern Counties Railway, to protect a 'Continuous Expansion Locomotive' developed by John Nicholson of the ECR. At least two locomotives (a 2-2-2 for passenger service and an 0-6-0 for goods traffic) were made on the Nicholson/Samuel system, but though they were said to have worked well enough, no more were ever made. The operation method was not a true compound, but was nonetheless a pointer to the future.

A three-cylinder compound 0-4-4-0 tank engine was proposed in the mid-1860s for suburban passenger train service by Jules-Émeric-Raoul Bricheteau de la Morandière (1837–1900), employed by the locomotive department of the Compagnie de Chemins de Fer du Nord. The British periodical *Engineering* illustrated the design on 23 November 1866, as 'Proposed Three-Cylinder Locomotive', but there is no evidence that it was ever put into traffic. A single high-pressure cylinder between the frames drove on to a crank built into the leading axle of the rear power unit, then exhausted to two low-pressure cylinders mounted outside the frame ahead of the leading wheels. These drove the rear wheels of the front power unit through conventional connecting rods. Both sets of axles were coupled, but the two power units were separated within rigid frames.

The first locomotive to incorporate true compound operation is believed to have been Erie Railroad 4-4-0 *No. 122*, delivered from the Boston Locomotive Works in 1851 but adapted to accord with US Patent 70341,

'Improvement in Railway Locomotives', granted on 29 October 1867 to John L. Lay of Buffalo, New York. Unfortunately, No. 122 was also fated to disappear into obscurity, as compounding clearly did not improve performance sufficiently for other Erie locomotives to be adapted.

John Lay, born in 1832, was listed in the 1868 Buffalo city directory as 'engineer'. He had served in the Federal navy during the Civil War, designing the spar torpedo that sank CSS *Albemarle*. Lay was eventually renowned as one of the fathers of the self-propelled or locomotive torpedo. However, a career in which he worked for several years in Russia, lost a fortune and wrecked his marriage ended in penury in a New York hospital in 1899.

The first truly successful embodiment of compounding in a railway engine occurred when Jules-Théodore-Anatole Mallet (also renowned for the articulated locomotives associated with his name), to whom patents had been granted in 1874, built his first locomotive in 1876: inspired, no doubt, by the success of Woolf-type compound stationary engines in France. The three 0-4-2Ts delivered to the Bayonne & Biarritz Railway embodied what was to become one of the standards: a two-cylinder system with one high-pressure and one low pressure cylinder connected to a common axle. The tank engines had cylinders outside the frames, but there was no reason why, if space permitted, they could not be accommodated internally. Inside cylinders had several advantages, not least minimal loss of heat and a reduction in the 'hunting' or oscillation that could accompany engines equipped only with outside cylinders.

The principal advantage of the Mallet system was the inclusion of a starting or 'simpling' valve, which allowed the driver to admit steam at boiler pressure to the low-pressure cylinder. This overcame the tendency of compounds to start sluggishly, relying only on the steam admitted to the high-pressure cylinder to overcome the inertia of the low-pressure piston. In addition, the Mallet valve allowed the locomotive to revert at any time to simple expansion if excessive demands were made by a severe gradient or a heavy trailing load. This was customarily known as 'reinforced working', but demanded some skill on the driver's part. It was most popular in France, where *mécaniciens* – and, indeed, locomotive engineers – enjoyed a status unmatched in Britain, where work on the railways was regarded as only one step forward from blacksmithing. Even in Germany, and also usually (but not inevitably) in the USA, simpling valves usually disengaged automatically once the driving wheels had made a few turns. This provides the principal difference between the manually controlled Mallet and the automated von Borries systems.

In Britain, the first experience of this type of operation was less than happy: it dealt compounding a blow from which it never recovered. The villain of the piece was Francis Webb, a skilful if dictatorial engineer who, by the early 1880s, had become locomotive superintendent of the London & North Western Railway.

Aware of developments taking place in France, Webb determined to build comparable engines in Britain. The LNWR was essentially a 'small engine' company, but a steady growth in train weight was taxing the existing simple expansion 2-4-0 locomotives to their limit. However, the LNWR was notoriously parsimonious; its rolling stock was plain, livery was comparatively simple (rivals were often a riot of colour), and money would not be spent unnecessarily. This limited what could be done. Webb's compound locomotives received a particularly bad press, but their designer had to work within restrictions that did not apply elsewhere. One was provided by minimal clearances between the track and LNWR platform edges, which restricted the width of rolling stock bodies. Another was a universal fear that if the centre line of a locomotive boiler exceeded a particular height above the rails – usually reckoned to be 7ft 6in – then stability would be compromised and trains would fly off the track. And few engineers, even assuming they accepted the value of coupled wheels, were prepared to fit connecting rods that exceeded the accepted maximum length ... which was also supposedly 7ft 6in! These factors conspired to ruin the reputation of compounding in British railway circles almost before it had had time to grow.

There was some justification for Webb's caution. The iron used to make the wheel tyres of the 1880s was often surprisingly soft, and varied from batch to batch. If coupled wheels were driven from a single axle, the drive wheel took most of the piston thrust on its crank or crankpin. Fitting a coupling rod merely assured that the accompanying wheels revolved at precisely the same rate. However, the drive wheel customarily wore more rapidly than its companions; eventually, the wheels, which differed slightly in diameter, tried to rotate at different speeds. When this occurred, stresses and strains could break the artificial constraint (the connecting rod). There had been well-publicised instances of rods failing long before Webb began

One of many 4-2÷2-0 compounds designed by Francis Webb, *Teutonic* was built in the Crewe workshops of the London & North-Western Railway. These locomotives were a brave effort to introduce sophistication to an English railway, but were hamstrung by the autocratic nature of their designer. They had soon either been consigned to the scrapheap or rebuilt by Webb's successor in simple-expansion form. (From an engraving published in the British periodical *Engineering*)

his work, and there were designers, James Stirling among them, who maintained that only large-diameter single drivers should be fitted to locomotives intended for high-speed passenger trains.

In the late 1890s, a trial had been undertaken in Germany with a four-cylinder compound 4-4-0 to investigate the popular view that coupling rods hindered performance. The locomotive, sitting 'dead' with no fire, was allowed to run freely down a 1-in-200 gradient to a checkpoint, at which its speed could be assessed. The first trial gave 24.8mph as a 4-4-0; when the connecting rods were detached, the resulting 4-2-2 reached 31.7mph. This demonstrated clearly that connecting rods *did* act as an inhibitor.

Webb began his work in 1878, when he converted a 2-2-2 designed by his predecessor Francis Trevithick by inserting a liner in one of the cylinders and laying a pipe or 'receiver' from what had become the high-pressure unit across the frames to connect with the other original cylinder, unaltered, which became the low-pressure side. Extensive testing showed that the changes were beneficial; though power had been reduced, the engine ran freely and economically.

Emboldened, Webb built the three-cylinder 2-4-0 *Experiment* in 1882, with two high-pressure cylinders outside the frames and a single large low-pressure unit inside. Hard running showed that the engine worked satisfactorily, so nineteen examples of a modified version were made in the Crewe manufactory. These soon proved too light to compare with existing two-cylinder simple expansion engines and were supplemented with the larger 'Dreadnoughts' of 1884, which weighed 42½ tons in working order.

From this basis, Webb applied compounding to everything from small tank engines to 0-8-0 four-cylinder freight locomotives. They have received a universally bad press, owing to a combination of badly proportioned cylinders and inefficient valve gear, and the lack of coupling rods on the earlier express locomotives was also a hindrance. The Teutonics had slip eccentrics on the inside cylinder and could revolve both sets of wheels independently, if not always in the same direction. This occasionally happened when the driver had backed on to his train. Even though Webb provided high-pressure cylinders that were larger than the optimal size, this was still often not enough to overcome the inertia of the low-pressure piston at 'right away'.

It is arguable if the Webb compounds were particularly economical. Typical of the barrage of statistics produced by the LNWR in an attempt to silence criticism were those published in *Engineering* on 11 May 1894, which gave the distances run and average coal consumption for all the Webb compounds from the date of 'first turning out' to the end of December 1893. The ten Teutonics had run 2,357,759 miles, the best performer (as far as coal consumption was concerned) being *Adriatic*, turned out on 19 July 1890, which had run 219,760 miles at an average consumption of 31.3lb 'per mile'. At the other end of the scale was *Oceanic*, turned out on 27 June 1889, which had run 299,816 miles at 37.9lb of coal per mile.

Even though the figures allowed 1.2lb coal for 'raising steam', they do not seem to be very impressive. Loads behind the engines were rarely excessive, and the best simple expansion engines running on other lines were demonstrably more economical. Unfortunately, arbiters such as 'coal consumed per train mile' are impossible to assess if the weight of the train 'behind the drawbar' and the conditions under which trials had been undertaken are unknown.

Webb compounds, lacking starting valves or the complicated controls of their French contemporaries, were comparatively simple and could run long distances between overhauls; they often kept to time much better than their fiercest critics have allowed, but they were still primitive compared with the de Glehn/du Bousquet 4-4-2s of the Chemins de Fer du Nord.

The last Webb compounds had four cylinders placed in a line across the frames to drive on a single axle. They performed well in service, though the original valve gear was not as good as it could have been. Beneficial changes were soon made and performance improved still further when Webb's successor, his one-time assistant George Whale, made additional modifications. But Whale and Webb had had an uneasy relationship, and the three-cylinder compounds were being dispatched to the scrapheap as soon as Webb had resigned his post in 1903.

By this time, however, train weight had grown past the point at which even the most powerful and sophisticated single drivers could pull them. Connecting rods had finally become indispensable, and, fortunately, hard iron and steel tyres were solving problems of differential wheel rim wear … but that was little consolation as far as Webb was concerned.

Webb compounds, usually of 'Teutonic' 2-2÷2-0 type, made by Dübs or Sharp, Stewart & Company, were tested by the Chemins de fer de l'Ouest in France, by the Pennsylvania Railroad in the USA, in Argentina, in Austria and in Brazil. Robert Stephenson & Co. Ltd even made two 4-2÷4-2 Webb-type tank engines for the narrow-gauge Antofagasta Railway in Chile. However, only the Oudh & Rohilkhand Railway in India purchased 2-2÷2-0 passenger engines in quantity, ten being supplied by Dübs in 1884–85.

These engines were not efficient enough to persuade even the Oudh & Rohilkhand to abandon simple expansion for compounding, and the railway, which was privately owned (if backed by government guarantee), lost its independence before the locomotives could demonstrate any potential they may have had. Yet the locomotive superintendent of the Scinde, Punjab & Delhi Railway, Charles Edward Sandiford, prepared drawings for two- and four-cylinder compounds of his own design as early as 1883. The four-cylinder version, with two high-pressure and two low-pressure cylinders driving a common axle, is said to have been essentially similar to the 1886 de Glehn design.

Sandiford converted a 2-4-0 to each of his designs, sending the two-cylinder *Vampire* and four-cylinder *Vulcan* for trials in the summer of 1884. *Vulcan* proved to be the better steamer and rode much better than the *Vampire*, but enginemen preferred the latter as it was easier to drive. Savings in coal consumption of 10–12 per cent were praiseworthy, particularly as the compounds regularly took heavier trains than the simple expansion 2-4-0s, but changes in the structure of the railway brought the experiment to a halt.

Elsewhere, compounding proved to be much more successful. In 1880, August von Borries had designed a two-cylinder compound system with an automatic starting valve, which overcame the worst problem of the earliest Webb locomotives. Two small tank engines had been tried on the Hannover division of the Prussian state railways, proving successful enough for a large goods engine to be altered in 1882. The two-cylinder system had soon been extended to a wide range of other types.

In 1884, Thomas Worsdell, then serving the Great Eastern Railway as locomotive superintendent, applied a modification of the von Borries system to two GER 4-4-0 passenger engines. No sooner had this been done, however, than Worsdell moved to the North Eastern Railway. There he developed a range of compounds with their cylinders inside the frames, though the steam chests were often placed outside. The goals were increased thermal efficiency and a reduction of stress on the moving parts. British Patent 999/85 was obtained on 23 January 1885, and, between 1886 and 1892, more than 200 Worsdell–von Borries compounds were made; most were 0-6-0 goods engines, but there were also twenty high-speed 4-2-2s. Though not without constructional problems, the engines were surprisingly efficient.

Thomas Wilson Worsdell was born in Liverpool on 14 January 1838, the son of Quaker coachbuilder Nathaniel Worsdell and Mary Wilson. He is best known for his tenure of the locomotive superintendency of the NER, where he introduced several types of two-cylinder compound locomotive based on the designs of von Borries. Worsdell retired in 1890, but continued to work for many years in collaboration with von Borries and Richard Lapage. Worsdell died in Arnside, Westmorland, on 28 June 1916. He and his wife, Mary Ann Batt, had several children, one of whom, Henry, also became involved in engineering.

Unfortunately, the external valve chests tended to crack and flaws in the design of the valve gear inhibited performance; Thomas Worsdell's successor, his younger brother Wilson, had soon converted most of them to simple expansion. Yet this was very rarely true of the thousands of von Borries compounds built in Europe, where, perhaps, a more scientific approach to railway engineering than was customary in Britain prior to 1914 had paid dividends.

Thomas Worsdell continued to improve his locomotives, obtaining patents in collusion with von Borries and a civil engineer, Richard Herbert Lapage. These designs had no influence on British railways, but were incorporated in many engines exported from Britain for service in the colonies and elsewhere.

The rapid advance of compounds in Europe had its origins in France. The work of Anatole Mallet has already been highlighted, but among the first to introduce compounding to the French mainline railway companies were Alfred de Glehn, Gaston du Bousquet and Adolphe Henry. With a few honourable exceptions, English language sources have been keen to credit de Glehn as the initiator of what had soon become a largely French phenomenon prior to 1914: compounding virtually everything from the smallest tank engine to the largest express passenger locomotive. This is perhaps due to the perception of de Glehn as English; to the mistaken assumption that only Britons contributed to the development of the locomotive prior to 1900; and perhaps also to attempts to counterbalance damage widely perceived to have been done to the reputation of compounding by the much-maligned Webb. De Glehn was born in England, in Marylebone, but was a true European; du Bousquet was fortunate to be able to improve de Glehn's prototype; Henry was fated to die before the value of the work he had started could be widely recognised. To whom due credit should be given, however, is an open question. All three men owed a debt to Mallet, who, like so many railway engineers, was willing to share his work by publishing articles regularly in engineering journals.

According to Édouard Sauvage, plans had been prepared in 1884 and what was to become locomotive No. 701 of the Compagnie de Chemins de Fer du Nord was delivered from Société Alsacienne de Constructions Mécaniques in 1885. Designed by Alfred de Glehn of SACM, the new engine was deemed to be such a radical departure from Nord practice that it was offered 'at manufacturer's risk': if it failed to work properly, the railway could return it without paying. Two high-pressure cylinders between the frames drove the leading axle, and two larger low-pressure units outside drove the rear axle. An inverted form of Walschaerts' gear was used, but, at the beginning at least, there was no simpling valve. Ironically in view of the lack of success of the Webb compounds, de Glehn also declined to couple the driving wheels – for precisely the same reasons that Francis Webb was giving in Britain.

No. 701 was displayed in Paris in 1889, but protracted trials had already shown that although the engine performed well enough to confirm many of the advantages of compounding, independently driven axles caused more problems than they solved. In 1891, therefore, two locomotives developed by du Bousquet from No. 701 entered service. Numbered 2121 and 2122, they were substantially larger than their prototype, weighing 47.8 tonnes instead of 37.8, and had coupled driving axles. A simpling valve system permitted the driver to choose two-cylinder simple expansion (using either set of cylinders), four-cylinder simple expansion, or 'reinforced compound' operation where some high-pressure steam could be admitted to the low-pressure receivers to boost power. In addition, though divided drive was retained, du Bousquet reversed the layout so that the low-pressure cylinders lay within the frames. Trials soon showed that the new locomotives exceeded expectations, and virtually every subsequent Nord 4-4-2, 4-6-0 and 4-6-2 design was a four-cylinder compound. A typical saturated compound 4-4-2 of the early 1900s, weighing only 65 tonnes, could take loads of 300–350 tonnes at 120kph (75mph) and indicate 1,400hp at these speeds.

The lesser-known story of the compounds of the Chemins de fer de Paris à Lyon et à la Méditerranée ('PLM') is comparable with that of the Nord. Adolphe Henry (1846–92) is said to have begun work on them as early as 1883, citing his debt to Anatole Mallet. It is assumed that experiments were undertaken by converting an obsolete locomotive, but the first to be made specifically in the railway's workshops in Oullins, a suburb of Lyon, were C1 and C2 (2-4-2) and 4301 and 4302 (0-8-0). One of each type was exhibited in Paris in 1889. The 2-4-2 passenger engine replicated the notation of an existing class of simple expansion machines, giving what was touted as a proper basis for comparison. The same was claimed for the 0-8-0, intended for fast freight and mixed-traffic roles. However, comparison was hindered by boiler pressure of 15at (213psi), one of the most remarkable features of the Henry compounds and judged to be almost dangerously high for its day. This alone skewed results greatly in favour of the compounds.

Both compounds had four cylinders, two high pressure between the frames and two low pressure mounted externally. The drive differed: the high-pressure cylinders within the frame of the 2-4-2 drove the front axle, and the low-pressure cylinders on the outside were connected by rods to crankpins on the rear wheels. The length of the connecting rod was constrained by an extra long piston rod, with the slide bars over the front driving wheel. The 0-8-0 relied on inclined high-pressure cylinders ahead of the front axle to drive a crank on the second; and the outside low-pressure cylinders, under the smokebox, were linked by rods to crankpins on the third pair of wheels. Another obvious difference could be seen in the firebox, cylindrical in the 2-4-2 and Belpaire-form in the 0-8-0.

Trials of the new PLM compounds showed they were economical to run, and could pull loads well in excess of those taken by simple expansion locomotives of the same notation (though the substantial increase in boiler pressure and greater overall weight had played their part). The decision to abandon the 2-4-2 was taken, and so three additional passenger engines were made in 1892: 2-4-0 C51 (later 'C3'), and 4-4-0 C11 and C12. Trials showed predictably that the leading truck was not as effectual as the bogie, and that the valve gear could be improved. An adaptation of Gooch link motion replaced Walschaerts' gear on the high-pressure cylinders of C11 and C12, and the pattern for future PLM passenger engines had been set – with the low-pressure cylinders inside the frame, between the bogie wheels, driving a crank on the leading axle; and each of the low-pressure cylinders, between the rear bogie wheel and the front driver, connecting with a crankpin on the second coupled wheel. An automatic simpling valve became standard.

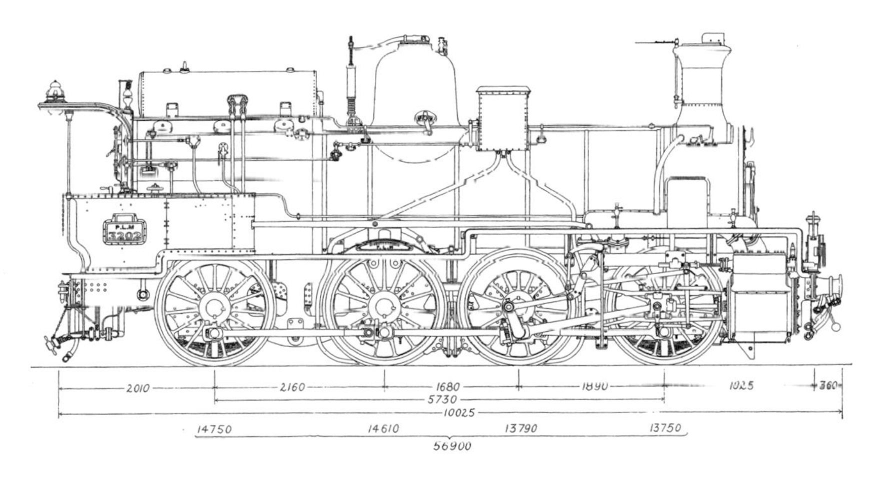

PLM 0-8-0 Henry four-cylinder compound No. 3202. Many locomotives of this type were subsequently converted by substituting a bogie for the leading coupled wheels. The resulting 4-6-0 looked ungainly, but ran much better than the 0-8-0 on fast mixed-traffic services.

There were also two improved 0-8-0s, 3201 and 3202, which had been under trial since 1890. In these the HP cylinders between the first and second axles drove a crank on the third axle, and the external low-pressure cylinders drove crankpins on the second set of coupled wheels. Reports made in 1892–93, after Henry had died, suggested that consumption of fuel compared with existing simple expansion 0-8-0s had dropped by an average of about 10.5 per cent. The best individual performance had saved 15 per cent. Consequently, the four-cylinder compound 0-8-0, once the boiler had been improved and a slight enlargement of the cylinders had been made, was accepted for service by Charles Baudry, Henry's successor.

Ever Onward!

By 1889, and the Exposition Universelle in Paris, the first steps towards an efficient railway engine had been taken. The principal operating characteristics had been established, though the way in which they were achieved differed from railway to railway, and among individual manufacturers. There were two basic classes of compounding: Woolf, in which steam flowed directly from the high-pressure to the low-pressure cylinder, and the 'receiver compound' in which a large reservoir was interposed to accept steam from the high-pressure cylinder before it could be admitted to the low-pressure cylinder.

The Woolf system was restricted to two cylinders between the frames, or paired cylinders outside (or, alternatively, one cylinder of the pair on each side of the frame plate). Receiver compounds, conversely, could use any number of cylinders as the steam supply was not directly from one cylinder to another. This was ideally suited to three-cylinder systems in which a single, central, high-pressure cylinder was flanked by two comparatively small-diameter, low-pressure units.

If there were only two basic methods of compounding, the ways in which they were achieved were much more numerous. Most cylinders drove their own individual crosshead and drive rod, connecting either to a crank on an axle (usually internal) or a crankpin on the driving wheel. Locomotives could have two cylinders (HP and LP) between the frames, or one exposed on each side; three-cylinder designs had one cylinder between the frame and one on each side (usually one central HP and two LP, though Webb compounds had two HP cylinders outside and one large central LP cylinder). Four-cylinder engines could have two cylinders between the frame and two outside, but tandem compounds usually had all four cylinders mounted externally.

The basic tandem idea had a lengthy pedigree, but, as far as commercial exploitation, was confined largely to the Nord railway of France and the Brooks Locomotive Company of Dunkirk, New York. Only three tandem compounds ever ran in Britain: a Nisbet patent locomotive on the North British Railway, converted from the Wheatley 4-4-0 that had been lost in the Tay Bridge disaster, and two engines, No. 7 and No. 8, designed by William Dean for the Great Western Railway. The NBR locomotive seems to have worked satisfactorily, GWR No. 7 was put briefly into main-line service, but GWR No. 8 was a disaster – blowing the mechanism apart three times in quick succession – and was abandoned immediately.

The Brooks Locomotive Company made tandem compounds in accord with US Patent 499584 granted to John Player on 13 June 1893. The cylinders were made in one compact unit, with a single wall forming the rear face of the leading high-pressure cylinder and the front face of the trailing low-pressure cylinder. A single piston rod and associated connecting rod were used. However, the otherwise generally comparable tandem compound system employed by Gaston du Bousquet on the large-wheeled 4-6-0 tank engines made for the Ceinture railway connecting the principal Paris railway termini, had the cylinders spaced sufficiently widely that the trailing bogie wheel passed between them. Construction of this type required separate valves for each cylinder.

The first Ceinture locomotives were criticised harshly by those who believed that separating the cylinders would lead to leaking glands and other maladies arising from the frames flexing on the road. Service soon showed that the engines were fast, smooth running and easily maintained; several of them were to serve for more than fifty years.

The Baldwin Locomotive Company made large numbers of four-cylinder compounds on a superposed cylinder system patented by Samuel M. Vauclain in 1889. This was much simpler than the Player tandem compound, as the high-pressure cylinder was placed directly above the low-pressure unit, separated by the solitary valve. The two piston rods drove on to a common crosshead, giving Vauclain compounds a distinctive appearance. They were very successful when new, but power generated in the paired cylinders was difficult to balance and the valve gear began to wear as the locomotives aged. As this added to maintenance problems, even Baldwin admitted defeat; a much more conventional two-cylinder compound system was substituted in 1904, when the first example was built for the Plant System (although the locomotive proved to be too heavy and was sold almost immediately).

Attempts were also made to introduce what were called 'annular' or 'concentric' compounds, with cylinders inside each other. There were even triple-expansion versions of the annular system, with the small central high-pressure cylinder exhausting to the central mid-pressure unit, and finally to the outer low-pressure cylinder before exhausting to the atmosphere. None of these designs was successful. Not only did it prove difficult to arrange satisfactory connections with the drive axles, but the effects of heat loss (and, therefore, differential expansion) compromised the operation of the valves. A few Johnstone annular compounds – strange locomotives in more ways than one – ran in Mexico for a few years, but no comparable machines are known to have run successfully elsewhere.

Few of the railway engines exhibited from the London Exhibition of 1862 to the Paris *Exposition Universelle* of 1878 had shown originality, however much they differed in detail. Articulated locomotives were offered, particularly to haul heavy loads up steep gradients or around sharp curves, and some curious prototypes could be seen. Yet Britain was represented in Paris in 1889 only by a handful of commonplace designs: a 4-2-2 from the Midland Railway; *Onward*, a 4-4-0 of the South Eastern Railway; and the 'Gladstone' 0-4-2 *Edward Blount* of the London, Brighton & South Coast Railway. All three engines had inside cylinders. They were Victorian railway engineering at its best: solid, dependable enough, beautifully finished, pleasing to the eye, stunningly colourful, but technically uninspired.

A 2-6-0 goods engine designed for the Nord railway by Édouard Sauvage was also on view. Built in 1887, the engine had a central 46 × 50cm (18⅛ × 19⅔in) high-pressure cylinder and two 70 × 50cm (27⅝ × 19⅔in) outside cylinders driving the central driving axle. The low-pressure cranks were set at 90 degrees, with the high-pressure crank at 135 degrees to each of them. Trials returned a drawbar pull of 4,400kg (9,700lb), achieved while hauling a 540-tonne train up a gradient of 1 in 200, and 620 tonnes were taken up the same slope at 20kph (12½mph). However, owing to its small wheels, with a diameter of only 1.65m (5ft 5in), the Sauvage locomotive was not as influential as its design deserved to have been; it was not perpetuated, even though Sauvage became an influential teacher, a fluent champion of compounding and, therefore, Chapelon's mentor.

Things had not improved greatly by 1900. Angered by opposition in Europe to the Second South African War, and widespread support for the Boer cause, Britain had sent very few railway exhibits to Paris. This caused adverse comment throughout the English-language engineering press, though the tone was generally muted. However, the customary oddity could be seen in the 'cab forward' locomotive made by Schneider & Company of le Creusot.

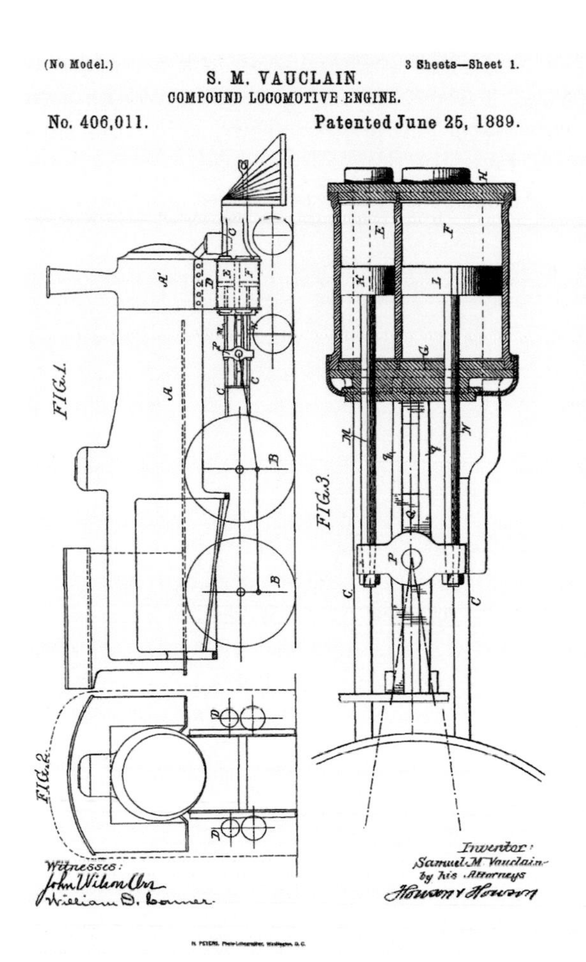

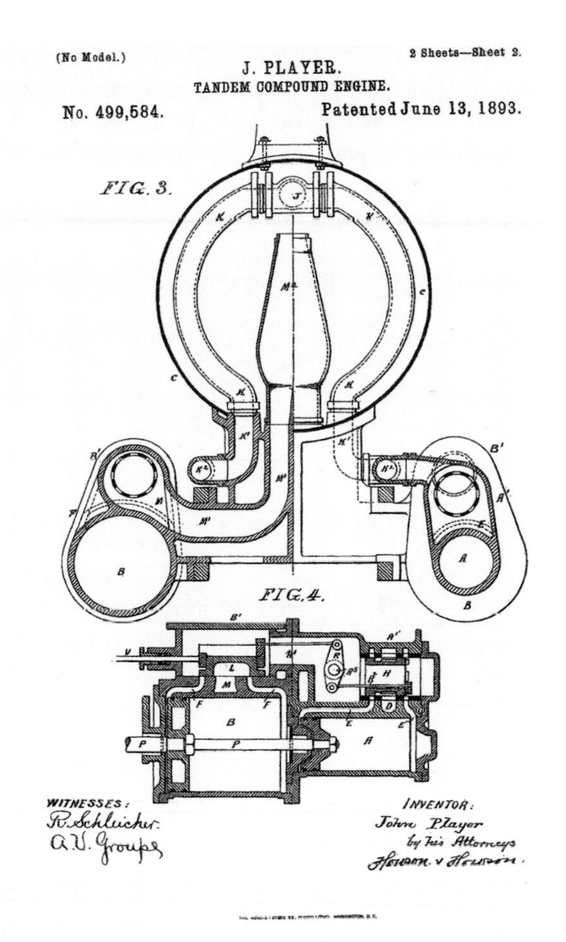

Left: Drawings from the patents granted in June 1889 to Samuel Vauclain of Baldwin (left) and in June 1893 to John Player of the Brooks Locomotive Works (right) to protect methods of compounding. Note that Vauclain's has the high pressure cylinder directly above the low-pressure unit, driving on to a common crosshead; Player's has the high pressure in front of the low-pressure cylinder, driving a common piston rod. (US Government Patent Office, Washington DC)

Below left: 0-4-2 No. 189 *Edward Blount* of the D1 class, designed by William Stroudley and built in the Brighton Works of the London, Brighton & South Coast Railway in March 1889. This locomotive was exhibited at the Paris Exposition Universelle, gaining a gold medal (suitably acknowledged on the rear splasher). No. 189 was withdrawn in December 1912 and scrapped.

Below: The unsuccessful supposedly high-speed Thuile 4-4-6 *Trains Internationaux No. 1*, built by Schneider & Cie and shown at the 1900 Paris Exposition. The locomotive is seen outside Chartres depot.

The machine had its origins in patents granted in the 1890s to Henri Thuile and Emmanuel-Jean-Baptiste Fouré, whom the British specifications identify as 'Engineer of the Port of Alexandria' and 'Customs Collector of Alexandria' respectively. Thuile had also been secretary to the king of Egypt. The original locomotive was a 6-4-6 with dual steam and electric drive, relying on a 'donkey engine' to drive a dynamo. This was omitted from the Schneider 4-4-6, which had two external cylinders, Walschaerts' gear, and wheels so large that the boiler had to be squeezed laterally to fit within them.

Thuile saw his *Trains Internationaux No. 1* as the motive power for a rival to the Orient Express. Unfortunately, trial runs between Chartres and Orléans revealed serious problems: the ride was unsteady, and, as too little of the considerable weight of the engine was available for adhesion, only lightweight trains could be pulled even though a maximum speed approaching 120kph had been attained. Thuile was killed in June 1900, apparently struck by a lineside pole when a sudden lurch threw him off the engine, and the project came to an ignominious end when the locomotive was scrapped in 1904.

French railways were particularly fond of compounding, which gave substantial economies when comparatively poor quality coal was used. This Serie 11 4-6-0 of the Chemins de fer de l'Est, No. 3112 dating from 1907, had two outside high-pressure cylinders exhausting into two low-pressure units within the frames.

The greatest exponents of compounding in this era were still French: perhaps fittingly, as the initial success of the compound stationary engine was gained largely in France. On 1 January 1900, the French railways' inventory of compounds included 350 of 4-4-0 or 4-4-2 type; 270 4-6-0s or 4-6-2s; and 176 0-8-0s. To put this success in context, British railways had between them nothing more than the surviving Webb locomotives of the LNWR and the Worsdell-von Borries patterns of the NER. The only British 4-6-0s were the goods engines that had been running on the Highland Railway since 1894, two aberrant outside frame prototypes of 1896 being tried on the Great Western Railway, and the first of a small class of NER passenger engines introduced in 1899. None of them were compounds.

The fast-growing success of the de Glehn-du Bousquet locomotives caused another look to be taken at compounding by the British. The Great Western Railway experimented in the early 1900s with three locomotives imported from France, before settling on four-cylinder simple expansion, but the use of English drivers may have been sufficient to prejudice the results of the running trials.

The best work in Britain was done by Walter Smith, chief draftsman of the North Eastern Railway, who modified a Worsdell simple expansion 4-4-0 to a three-cylinder compound system in 1898. This had a central high-pressure cylinder exhausting to two external low-pressure units, with the 90-degree/135-degree crank settings pioneered by Édouard Sauvage on the Nord railway in France. Separate control gear was fitted and high-pressure steam could be admitted to the low-pressure cylinders to start the engine away.

This locomotive remained a prototype on the NER, though well regarded. It inspired a handful of three-cylinder compound Atlantics on the Great Central Railway and the Smith-Johnson 4-4-0s on the Midland Railway, the first of which dated from 1902. Richard Deeley then built substantial numbers of a modified pattern – the 'Midland Compound' – after he had succeeded Johnson in 1906. These were popular and smooth-running engines, but were handicapped above 60mph by poor steam passage design.

Smith's final design, completed in 1904, was a four-cylinder compound Atlantic with two high-pressure cylinders inside the frames exhausting to two external low-pressure units. Two examples of this first-class design were built for the North Eastern Railway in 1906, one with Stephenson valve gear and the other with Walschaerts' gear, but the sudden death of Smith brought work to an end. The NER wanted to build more of them, but the mercenary attitude of Smith's executors forced the substitution of heavy three-cylinder simple expansion locomotives developing comparable power but a much greater 'hammer-blow' on the track. Consumption of coal and water also increased appreciably.

Walter Mackersie Smith was an interesting man. Born on Christmas Day 1842 in Ferryport on Craig, Fife, he was lodging in Glasgow as a 'pattern maker' by 1861. Married to Margaret 'Maggie' Smith, he moved to England – working as an 'engineer' for the GER at Stratford in east London, and then as a 'locomotive engineer' for the NER in Newcastle upon Tyne. Smith's passion was compounding, and he spent much of his later career developing three- and four-cylinder designs. The success of the Midland Compound 4-4-0 owed much to Smith's early work, but the perfected four-cylinder NER 4-4-2 had hardly demonstrated its efficiency when he died in Newcastle in October 1906. His executors kept such a stranglehold on Smith's many patents that the NER and others soon lost interest. Smith had eight children, two of whom, John and Walter, pursued careers in engineering.

Winners and Losers

What was obvious from Paris in 1900, however, was that Britain was falling behind France and even Germany in the design of railway engines.

The rapid spread of railways kept British manufacturing companies in constant work until the First World War, and ensured that a sizeable worldwide export trade in railway equipment had been created. Yet the mass emigration to North America that took place during the nineteenth century, especially from the British viewpoint, posed a very real threat: the rapid establishment of competition.

Many of the advances in engineering that occurred in the USA were due to skilled immigrants from central Europe, and the emergence of industries in India and Australia also began to threaten British colonial trade. By the beginning of the twentieth century, the Lancashire cotton industry was threatened by nearly 200 cotton mills operating in British India, mostly in and around Bombay; daily employment was estimated at 156,000 men, women and children. Though the first Indian mill had been opened only in 1851, capacity had more than tripled in the 1880–1900 period. And even then, it was said, 'the products of the Japanese mills are rapidly cutting out those of the Indian mills in the Far Eastern markets …'

By this time, Canadian railways had grown to a 'main-line' track length of more than 18,000 miles, owned by 163 separate operators, and had carried almost 37 million tons of freight in the year to 30 June 1901. Canada had never been a strong market for British locomotives, rolling stock or rails – the proximity of the USA was largely responsible – but the situation elsewhere soon became critical. It was quicker to ship to Australia and New Zealand from San Francisco than from Britain, even though the Suez Canal shortened the route that had once passed around the Cape of Good Hope. In addition, the emergence of Germany as a technological power, and to a lesser extent France and Belgium, began to erode markets in the Middle East and parts of Africa.

Against the backcloth of the Boer War, which did much to foster anti-British feeling in Europe, English language periodicals and newspapers began a campaign to discredit 'foreign goods'. The government even investigated the affairs of the railways in Egypt, then nominally a British protectorate, and published a White Paper (Cd. 1010) entitled

A Gölsdorf 4-4-2 108-Class compound of the Imperial and Royal (Austro-Hungarian) Railway, built in Prague in 1900.

Correspondence Respecting the Comparative Merits of British, Belgian, and American Locomotives in Egypt. The conclusion, predictably, was that British was best; however, protracted delivery of goods was criticised severely on the grounds that it 'left the door open' for foreign competitors.

A letter published in *Engineering* in 1901, from an engineer describing his experience on the 'Koyloff, Voroneja, Rostoff railway … worked chiefly with Belgian-made engines', expressed the popular xenophobic view:

> I may say at once that these proved a failure, the cylinders and smokeboxes got slack, the valve motion [was] unsatisfactory, and the material of a poor quality, particularly the boiler plates … About this time our goods traffic was completely blocked, and had it not been for two or three old Neilson engines [made in Glasgow] which worked the traffic from the coal mines to the port, I do not know what the consequences would have been … We also had about twenty American engines, which did good work, but gave us a lot of trouble with their boilers, fireboxes, and water-tube fire-bars, designed for burning anthracite. The workmanship of these engines was rough in the extreme, the wearing parts of the link motion much too light, and too much cast iron was used in their construction, instead of cast steel. Some Sigl (Vienna) engines we had did first-rate work, and their casehardening I have never seen excelled and seldom equalled.
>
> My next experience was on the Moscow Kursk Railway, where most of the engines were built by Borsig (Berlin); but for heavy work they could not be compared with the Beyer and Peacock engines, employed for the same service. I ought to say here, however, for the Borsig engines, that both the material and the workmanship were good, but that the design was too scientifically accurate; not a sufficient margin being allowed for wear and tear … I next moved to the Nicolai railway, where I had a still better opportunity of comparing British, German, American, and French-built engines, and I say again … that the British-built engines more than held their own, though the German engines – 'Kessler's' – were much the best I have seen from that country, and came nearest to our own …

It would be fair to say that the British locomotives of the 1870s were often superior to their foreign rivals, but also that this superiority, eroding fast by 1900, had all but disappeared by 1914. British industrial supremacy was clearly in decline; the USA and Germany had reduced Britain to third place in the production of pig iron in 1903, and Britain's output of manufactured goods, 32 per cent of the worldwide total in 1870, had dropped to merely 14 per cent when the First World War began. British locomotives, generally well built, were too often handicapped by aesthetic considerations: they looked good, with complex multi-coloured liveries, and were almost always acceptably reliable. But few British railways worried unduly about maintenance prior to 1914, as labour was plentiful and sufficient motive power existed to answer most traffic demands.

These conditions were not always paramount elsewhere, where ease of maintenance and sometimes also technical sophistication (especially in France) prevailed. American locomotives often looked ungainly to British eyes, and were rarely designed to have service lives of comparable longevity, but they were simple and easy to maintain. In addition, as long as the purchaser accepted the manufacturer's standards, locomotives built in the USA were usually supplied quickly and inexpensively.

In 1901, the ten principal British railway locomotive builders employed nearly 14,000 men, and tens of thousands were involved in the ancillary industries or trades. Three of the four largest companies worked in Glasgow: Neilson, Reid & Co., Dübs & Co., and Sharp, Stewart & Co. But even these three, when amalgamated in 1903 to form the North British Locomotive Co. Ltd, were unable to compete with the output of North American giants such as Baldwin or the American Locomotive Company ('Alco'), a 1901 amalgamation of several smaller US manufacturers.

The three components of NBL had made about 15,437 locomotives prior to the 1903 amalgamation; by the time the last order had been completed in 1958, NBL had made an additional 11,318. The Baldwin Locomotive Works built its 17,000th railway engine in September 1899; the 30,000th dated from February 1907; the 50,000th from September 1918; and about 60,000 had been made when production ended in the 1950s. By comparison, the overall total of locomotives built in Britain (a difficult figure to compute accurately) has been estimated at only 140,000. Most of the latter had been made in Glasgow, Manchester and the 'Railway Towns', such as Crewe or Swindon, which had sprung virtually from nothing in the middle of the nineteenth century.

It was realised from the earliest days that demands for rolling stock could be satisfied only by some type of standardisation – Matthias Baldwin, for example, was making engines with standard templates and gauges as early as 1838. Though engineers such as William Stroudley strove to introduce standard components and fittings, and though some nineteenth-century British classes contained hundreds of similar-looking engines (the Ramsbottom DX Goods of the LNWR eventually mustered 963), they were erected in batches and rarely truly interchangeable. This was largely due to the high degree of hand finishing applied during assembly.

After 1900, therefore, efficiently run locomotive manufacturers in the USA and Germany overhauled the British exporters by making huge numbers of locomotives that gained in ease of maintenance what they often lacked in style.

Standardisation enabled engines of the same type to be made by different manufacturers, instead of each company championing its own methods; it reduced prime costs, pioneered the organisation of sub-contractors and undoubtedly accelerated production. But, if pushed to extremes, it hamstrung independent thought and could be a 'dead hand' on development. The individuality of railway companies disappeared, standardisation being mirrored by the grouping of many smaller companies together, though quality and performance could be increased.

The quantities involved were often stupendous, but, almost without exception, locomotive classes standardised in this way were two-cylinder simple expansion machines offering great sturdiness but little in the way of sophistication. The Prussian approach was exemplified by the P8, a rugged simple expansion superheated 4-6-0 with 1.75m (5ft 9in) driving wheels. Designed to achieve 110kph (68mph), the 'universal locomotive' was intended to be used everywhere on the Prussian network by even the most inexperienced crews. However, poor balancing and unsteady riding, which limited speed to only 100kph (62.5mph), was never overcome satisfactorily.

The same problem was not as obvious in the comparable G8 0-8-0 and G10 0-10-0, used largely for heavy slow-speed goods trains, and was not enough to prevent staggering numbers of each design being made – at least 3,556 4-6-0 P8 (1906–23), alongside 6,858 0-8-0 G8, G8¹ and G8² (1902–28). There were also 2,615 0-10-0s of the G10 Class (1910–24).

The Russians and the Soviets built about 14,000 0-10-0s of near identical type, and there were other instances of large-scale production that could only have been possible by standardising components. However, problems arose when orders for huge numbers of locomotives were spread among

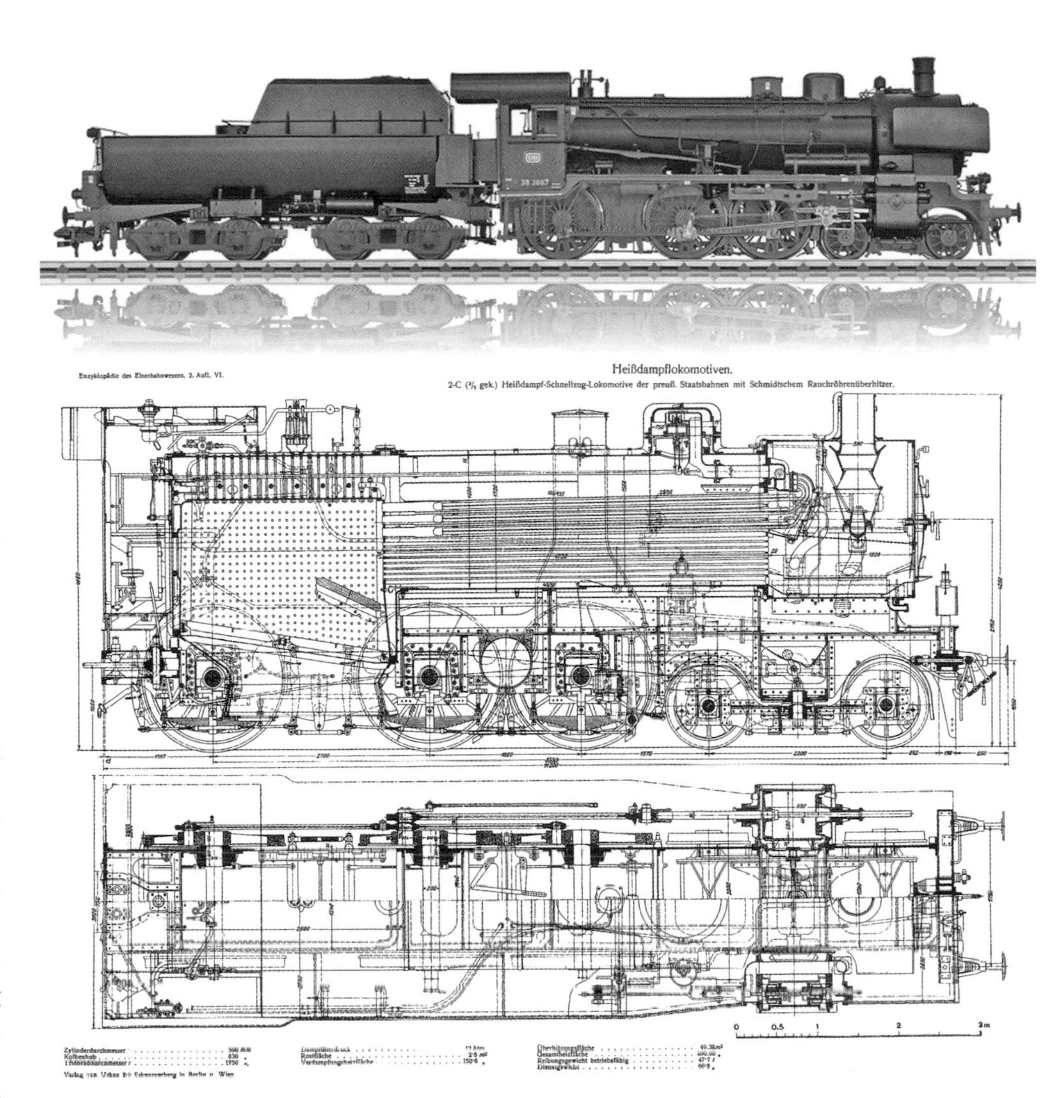

Drawings of the P8 4-6-0 of the Prussian state railways with a Schmidt superheater, from *Enzyklopädie des Eisenbahnwesens*, and a Märklin model showing alterations made after the end of the Second World War: a 'tub' tender, smoke deflectors and an extended chimney. (Model courtesy of Märklin)

THE BALDWIN LOCOMOTIVE WORKS, PHILADELPHIA, PA.

Six-Coupled Locomotives with Two-Wheeled Front Trucks for Freight Service
Mogul (2-6-0) Type

GAUGE	CODE WORD	Cylinders Diam. Stroke Inches	Diameter Driving Wheels Inches	Boiler Pressure Pounds per Square Inch	Rated Tractive Force Pounds	Weight in Working Order Pounds		Wheel Base		Capacity Tender for Water 8⅓-lb. gallons	Load in Tons (2000 Pounds) of Cars and Lading				
						On all Driving Wheels	Total	Of Driving Wheels	Total		On a Level	On a Grade per Mile of 26.4 ft. or ½ %	52.8 ft. or 1 %	105.6 ft. or 2 %	158.4 ft. or 3 %
3 Ft. or 1 Metre	Mattgruen ...	11x16	33	160	7,970	34,000	40,000	8′ 4″	14′ 4″	1500	815	380	230	120	75
	Matthanias ..	12x16	33	160	9,490	38,000	47,000	8′ 8″	14′ 8″	1800	960	450	275	140	85
	Mattheit	13x18	37	160	11,170	45,000	54.000	9′ 4″	15′ 6″	2000	1140	540	330	170	105
	Mattherzig ..	14x18	37	160	12,970	52,000	62,000	9′ 8″	15′ 11″	2200	1320	625	380	200	125
	Mattiaker ...	15x18	37	160	14,880	60,000	70,000	10′ 0″	16′ 6″	2500	1525	720	440	230	145
	Mattiamo ...	16x20	42	160	16,580	67,000	76,000	11′ 0″	17′ 10″	2800	1700	805	495	260	165
	Mattigkeit ...	17x20	42	160	18.720	76.000	86,000	11′ 6″	18′ 11″	3000	1940	920	565	295	190
4 Ft. 8½ Ins.	Mattinando ..	16x24	54	180	17,400	75,000	90,000	11′ 10″	19′ 7″	3000	1880	890	545	285	180
	Mattinassi ...	17x24	54	180	19.650	84.000	100,000	11′ 10″	19′ 7″	3500	2125	1005	610	320	200
	Mattinava ...	18x24	54	180	22,040	93,000	109,000	11′ 6″	19′ 4″	4000	2385	1125	690	365	230
	Mattinerai ...	19x24	54	180	24,550	104.000	121,000	12′ 0″	20′ 3″	4000	2660	1260	775	410	260
	Mattiola	20x24	56	180	26,220	115,000	132,000	13′ 0″	21′ 4″	5000	2835	1340	820	430	265
	Mattithiah ...	20x26	56	200	31.570	130 000	147,000	14′ 0″	22′ 5″	5500	3310	1570	960	505	320
	Mattock	21x28	62	200	33.850	144.000	164,000	14′ 6″	23′ 4″	6000	3690	1735	1065	560	355

Six-Coupled Locomotives with Four-Wheeled Front Trucks for Passenger or Freight Service
Ten-Wheeled (4-6-0) Type

GAUGE	CODE WORD	Cylinders Diam. Stroke Inches	Diameter Driving Wheels Inches	Boiler Pressure Pounds per Square Inch	Rated Tractive Force Pounds	Weight in Working Order Pounds		Wheel Base		Capacity Tender for Water 8⅓-lb. gallons	Load in Tons (2000 Pounds) of Cars and Lading				
						On all Driving Wheels	Total	Of Driving Wheels	Total		On a Level	On a Grade per Mile of 26.4 ft. or ½ %	52.8 ft. or 1 %	105.6 ft. or 2 %	158.4 ft. or 3 %
3 Ft. or 1 Metre	Matularum ..	12x18	42	160	8,390	38,000	51,000	9′ 6″	17′ 8″	1800	900	420	250	125	75
	Matulis	13x18	42	160	9,850	45,000	58,000	9′ 6″	17′ 8″	2000	1055	495	300	155	90
	Maturabunt ..	14x20	44	160	12,120	53,000	68,000	10′ 11″	19′ 5″	2200	1305	615	375	195	120
	Maturacao ..	15x20	44	160	13,900	59,000	76,000	11′ 6″	21′ 3″	2500	1500	705	430	225	140
	Maturamus ..	16x20	44	160	15,820	65,000	83,000	12′ 6″	21′ 10″	2800	1655	780	475	245	155
	Maturandos .	17x20	44	160	17,870	72,000	94,000	12′ 6″	21′ 10″	3000	1825	860	525	275	170
4 Ft. 8½ Ins.	Matureness ..	16x24	56	180	16,780	75,000	97,000	10′ 0″	20′ 1¼″	3000	1810	850	520	270	165
	Maturitaet ...	18x24	56	180	21,240	95,000	121,000	11′ 4″	21′ 7″	4000	2300	1080	655	340	210
	Maturorum ..	19x26	62	180	23,160	103,000	136,000	13′ 4″	24′ 6″	4500	2510	1175	715	370	225
	Maturuerat ..	20x26	62	180	25,670	117,000	150,000	13′ 6″	24′ 6″	5000	2760	1295	785	405	245
	Matuteabas ..	21x26	62	180	28,300	130,000	165,000	14′ 10″	25′ 10″	5000	3050	1435	875	450	280
	Matutearan ..	22x28	62	200	37,160	160,000	203,000	13′ 10″	25′ 10″	6000	4020	1900	1165	595	385

A Baldwin catalogue page showing a 2-6-0 Mogul and a 4-6-0 Ten-Wheeler – two of the standardised designs – and the configurations in which they could be supplied 'from stock'.

different manufacturers. Each manufacturer had long-established and often unique working practices and the result was that supposedly standard locomotives had many components that would not interchange with those of another make. True standardisation, therefore, was rarely achieved.

Though nearly 4,000 Prussian P. 8 4-6-0s were made in 1908–30 (including 226 made under licence in Romania), the initial orders were given to twelve contractors. Schwarzkopff, subsequently known as Berliner Maschinenbau, made 1,025 of them; and another 742 were the work of Henschel. The 277 'Ten-Wheelers' of the French Chemins de Fer du Nord made between 1897 and 1913 were the work of eleven agencies; and the 842 'Black Five' 4-6-0s made for the LMS and British Railways (1934–51) involved five primary contractors – three railway-owned works (Crewe, Derby and Horwich), Armstrong Whitworth and the Vulcan Foundry. The order for the first fifteen Belgian Class I Pacifics was split between Société John Cockerill; Les Forges, Usines et Fonderies de Haine-Saint-Pierre; Ateliers de Construction de la Meuse; and la Division de Tubize des Ateliers Metallurgiques.

Articulation

Enlargement of rigid frame locomotives in a search for power had drawbacks. Either the diameter of the driving wheels had to be reduced to fit within a specific wheelbase, generally lowering the speed that could be attained without wearing the valve gear, or the wheelbase had to be extended so that the wheel diameter could remain unchanged. Extending the rigid wheelbase too far confined the locomotive to well-laid main-line track with gentle curves, and restricted its utility greatly. The provision of flangeless wheels and axles that could move a short distance laterally eased some of the problems, as did the provision of bogies and swivelling trucks. But they were not the entire answer, and articulation often seemed a better alternative.

Though Englishman Matthew Murray proposed a design unsuccessfully in the mid-1820s, the first articulated locomotive to be built was the work of Horatio Allen in 1831–32. This ran in the USA on the South Carolina Railroad and had two twin-boiler units connected by a central firebox. It had two cylinders and a theoretical 2-2=2-2 wheel notation, but was not especially successful in service.

In 1851, the year of the Great Exhibition held in London, a competition had been held in Austria to find a locomotive that could successfully work on a railway line over the Semmering Pass, where steep gradients and high altitude were widely believed to make locomotive haulage impossible. The competition produced three differing approaches to articulation: *Seraing*, built by Cockerill of Belgium; *Wiener Neustadt*, by Günther of Vienna; and *Bavaria*, by Maffei of Munich. The Belgian locomotive is generally regarded as the prototype of the Fairlie system, while the *Wiener Neustadt* pre-empted the Meyer.

The Mallet became by far the most popular of all the systems of articulation, though this was at least partly due to its espousal by the most influential railroads in the USA. Its history began with a French patent granted to Anatole Mallet in 1884. In 1887, the Decauville company realised that the Mallet system (then conceived as a compound) was an ideal way of providing a 60cm-gauge engine that could haul loads equal to its own weight on 8 per cent gradients (1-in-12.5), travel safely around curves with radii as sharp as 20m, yet have an axle loading of less than 3 tonnes. The Decauville 0-4=4-0T Mallet carried the front power unit on a Bissell truck, pivoted to the frame roughly beneath the midpoint of the boiler. Overall length was only 5.43m (17ft 8in), weight in working order being about 11.6 tonnes. High-pressure steam was supplied directly to the rear cylinders, then exhausted into a receiver from where it was taken to the front or low-pressure cylinders by a flexible pipe. Locomotives of this type gained renown on the 60cm-gauge 'Inner Circle' track laid at the Exposition Universelle, Paris, in 1889. More than 6 million passengers were carried without a single serious incident.

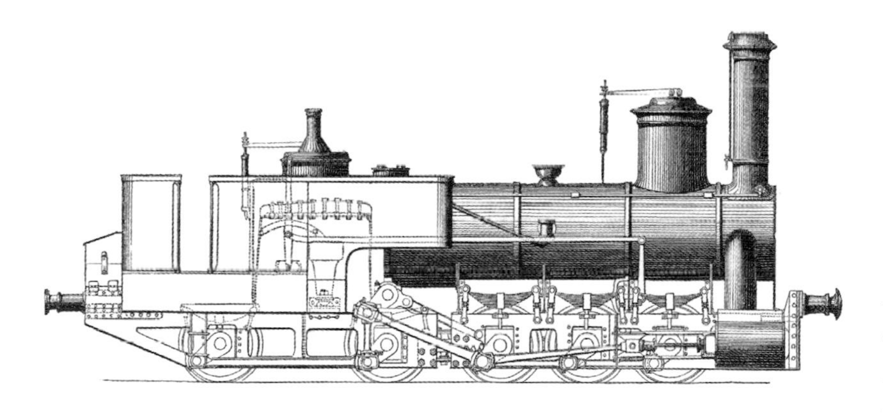

Designed by John Haswell, *Steierdorf* had a special Fink-pattern parallelogram linkage to transfer the drive from the third to the fourth axles, but was unsuccessful. (Official *Record of International Exhibition, 1862*)

Prior to about 1902, no notice of the Mallet had been taken in North America, where the heaviest freight traffic was being moved by large rigid frame locomotives such as 2-10-0s and 2-10-2s, 162 tandem compounds of the latter type being made for the Atchison, Topeka & Santa Fe Railroad in 1902–07. Unfortunately, the inflexible wheelbase and high axle loading of these engines confined them to well-laid track with shallow curves. In 1903, however, the American Locomotive Company of Schenectady built a 0-6=6-0 Mallet for the Baltimore & Ohio Railroad. The engine had two high-pressure cylinders measuring 20 × 32in, two low-pressure units of 32 × 32in, and 4ft 8in coupled wheels. It weighed 212½ tons with its tender.

Exhibited at the St Louis World's Fair in 1904, this Alco Mallet was the butt of much criticism from engineers, who were contemptuous of its great weight, the method of articulation and the design of the flexible steam pipes. When the Mallet entered service after the fair had closed, it proved much more successful than anyone excepting its promoters had anticipated. Minor problems were soon cured and a reputation for great power was gained. There followed an undignified scramble to develop the biggest and most powerful Mallet-type locomotives imaginable, the rivalry between the major US railroads being matched only by rivalry between Baldwin and the American Locomotive Company ('Alco'), formed by the amalgamation of several lesser lights in 1901.

The Mallet design also proved to be popular in Europe. By 1914, about 2,500 had been made by Decauville and Batignolles in France; by Borsig and Henschel in Germany; by the Hungarian state factory; by Putilov and Kolomna in Russia; and by the Swiss Locomotive Works in Winterthur. A few had even been built in Britain by the North British Locomotive Co. Ltd of Glasgow, beginning with four 0-6=6-0 engines made for China in 1907–09.

The largest of all conventional two-unit Mallets were the ten 2-10=10-2 examples built in 1918 for the Virginian Railroad by the American Locomotive Company. Intended to bank 15,000-ton freight trains over Clark's Gap, they lasted until the late 1940s. An obvious problem was posed by the length of the boiler on these locomotives; the front power truck was intended to pivot beneath the boiler, but concerns were raised that the lateral displacement of the smokebox could be enough to strike a train passing on a neighbouring track. Samuel Vauclain, realising that only the rear half of the boiler was instrumental in raising steam, patented a steel bellows to attach the two halves of the boiler shell in such a way that they could move with the curve! Needless to say, the idea was not successful.

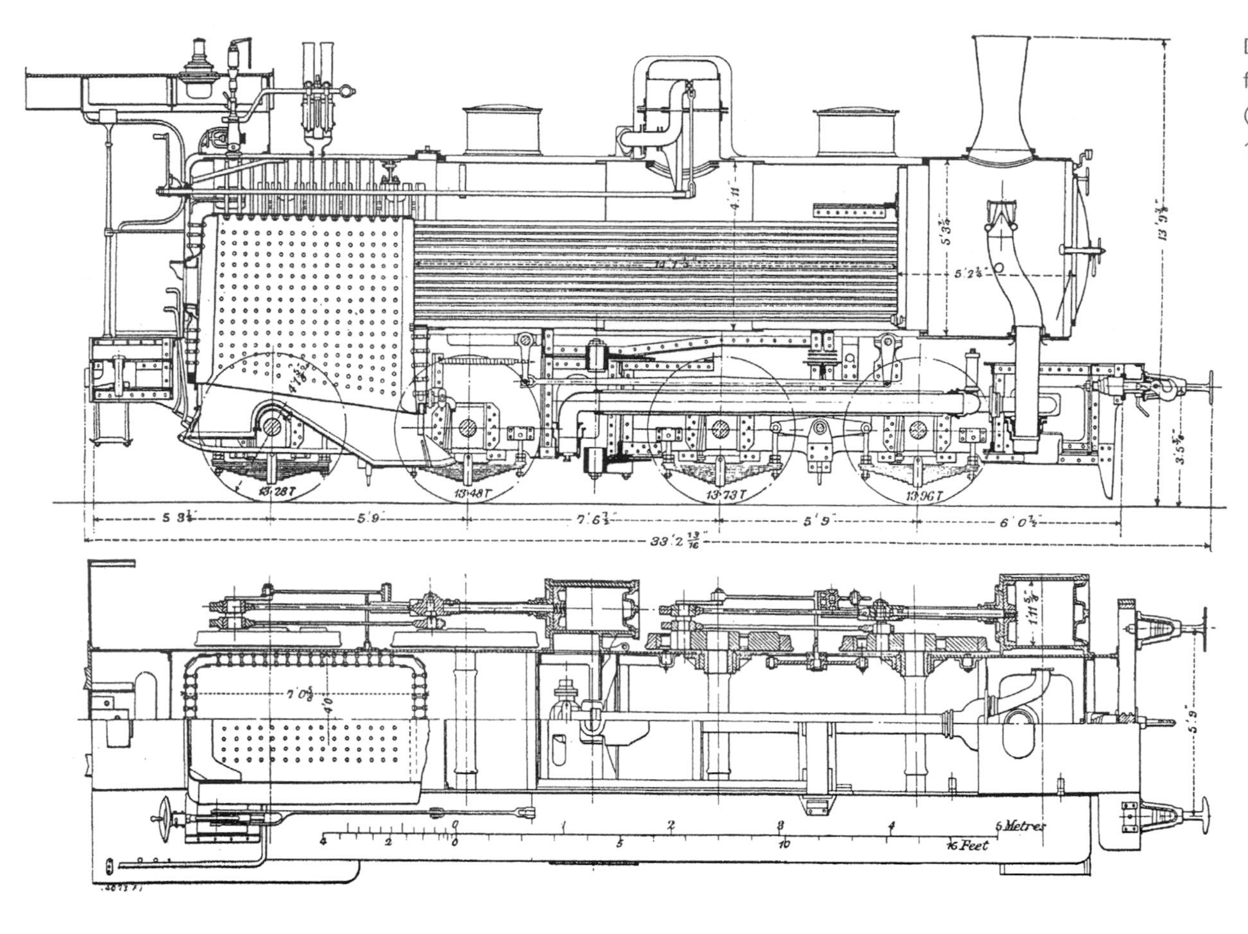

Drawings of the 0-4=4-0 Mallet made for the Prussian state railways from 1894. (British periodical *Engineering*, 10 July 1914)

Though some railroads retained faith in compound Mallets until the end of steam in North America – there was little doubt that they ran more economically than the simples – a majority preferred simplicity in the years after the end of the First World War. By this time the diameter of the low-pressure cylinders had grown so large that they became increasingly difficult to fit within even the generous US loading gauge.

An attempt was also made to perfect the so-called Triplex or Henderson Mallet with a fixed central power truck and articulated trucks at each end. Exhaust from the central high-pressure cylinders was split between two pairs of low-pressure cylinders positioned on the pivoting trucks. Though this enabled six identical cylinders to be fitted, the added complication and loss of heat from the attenuated steam pipes proved to be the undoing of the Triplex system. Only five locomotives seem to have been made, all by the Baldwin Locomotive Company. The prototype was *Matt H. Shay*, a 2-8=8=8-2 delivered to the Erie Railroad in 1913. Trials showed that tremendous loads could be hauled on level track (on one occasion, 16,300 tons were drawn at 14mph) but also that the boiler could not supply steam fast enough to reach higher speeds. This was partly due to a reduction in draught, as the rearmost cylinders exhausted to the atmosphere through an auxiliary chimney at the rear of the tender.

The most powerful railway locomotive of all time, on the basis of tractive effort, was 377-ton Baldwin-made 2-8=8=8-4T Triplex Mallet No. 700, dating from 1916. This had two high-pressure and four low-pressure 36 × 32in cylinders and a nominal tractive effort, with all

A stage too far. *Matt H. Shay*, a 2-8=8=8-2 Triplex Mallet of 1913. These gigantic locomotives offered huge power on paper, but practice showed that they did not steam well. Steam supply to the cylinders was difficult to balance, and loss of heat in the steam pipes was also a major problem.

cylinders working on simple expansion, of 199,560lb. However, the boiler would not supply sufficient steam at more than 10–15mph, so No. 700 was reconstructed as two separate engines in 1921. A design for a gigantic Quintuplex locomotive (2-8=8=8=8=8-2), with two cabs, is said to have been considered by Baldwin in 1916, but never left the drawing board. The problems of maintaining a good supply of steam, which had defeated even the 2-8=8=8-4 'Virginian' Triplex Mallet, would have presented even more of a problem with ten cylinders and five sets of driving gear!

Geared and Non-Standard Locomotives

Among the most interesting articulated designs was the geared drive locomotive patented in the USA on 14 June 1881 by Ephraim Shay (No. 242992). Intended specifically for use on logging railways and unique to the Americas, the first Shay had been built by the Lima Machine Company in 1880. The distinctive power train of a multi-cylinder inverted-vertical engine, driving the wheels through shafts, universal joints and gearing, reduced the axle load while ensuring that the entire engine weight was available for adhesion.

Shays were not only excellent hauliers, but also remarkable for their smooth running and the absence of the 'hammer blow' that arose from heavy reciprocating masses and rotating balance weights on conventional freight locomotives. A Shay with two power bogies was exhibited in Chicago in 1893 and a three-bogie 0-4-4-4-0 (works No. 867) at St Louis in 1904. The latter had three 12 × 15in cylinders, three power bogies with 3ft diameter wheels, and weighed about 58 tons in working order. Overall length was 47ft 2¼in.

Even though speeds were very low, the standard 0-4-4-4-0 could haul 2,800 tons on level track, 280 tons up a 1-in-30 gradient, and even 120 tons up 1-in-16. The minimum curve the bogies could take was just 100ft. Among the smallest pattern was the 2ft gauge '10-2 Shay', which

This two-unit Shay was built in December 1894. It was the only one to feature Joy valve gear and the first to have three cylinders. Note the bevel gear transmission. (US Forest Service)

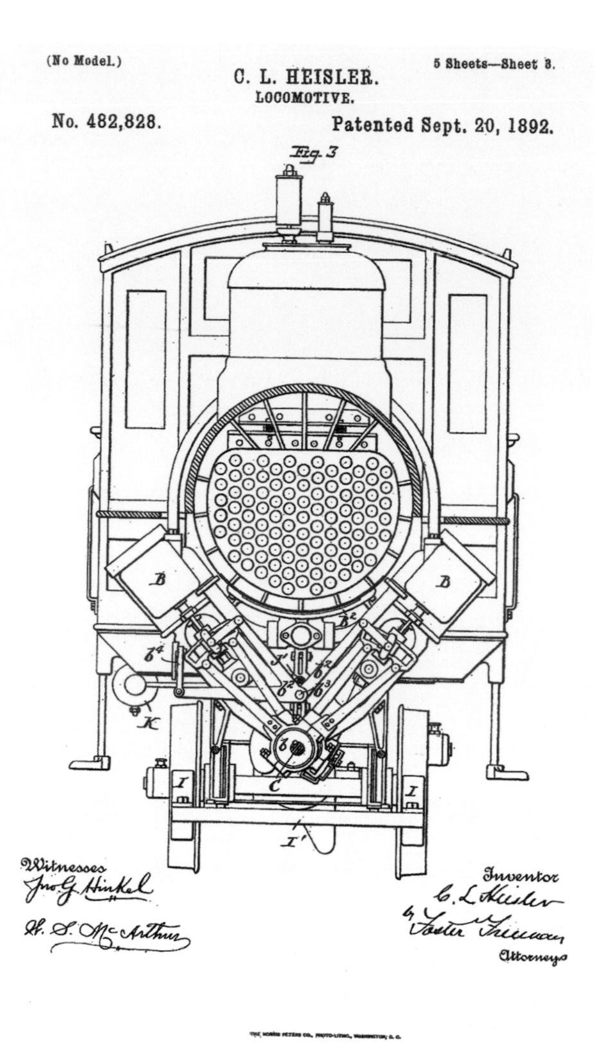

The Heisler locomotive had a 'V'-type engine driving directly on to a central transmission shaft. (US Government Patent Office, Washington DC)

had three 4½ × 8in cylinders. Though the diameter of the coupled wheels was only 22in, the engines could still pull 450–475 tons on level track.

About 2,770 Shays were made in Lima, Ohio, by the Lima Machine Works (1880–1931), and thirty-three of an adaptation of the basic design by the Willamette Iron & Steel Works of Portland, Oregon, in 1922–29. This made them the most popular of the geared locomotives manufactured in the USA. However, there were at least 1,045 Climax locomotives and more than 600 Heislers.

The Climax (which originally had vertical cylinders on the centre line) was designed by Charles Scott but protected by patents granted to George Gilbert of Corry, Pennsylvania, and Rush Battles of Girard, Pennsylvania. The prototype had been made under the supervision of Charles Scott, partner in the Scott & Akin Company, loggers working in Spartansburg, Pennsylvania. However, US Patent 393896 was granted on 4 December 1888 to civil engineer George Gilbert, who had transformed the Scott locomotive to suit it to series production. The two-speed gearbox proved to be unsatisfactory, so a fixed drive patented by Battles on 25 February 1890 (US No. 421894) was substituted. Longitudinal drive was replaced by transverse jackshaft and externally mounted cylinders in 1890.

Scott sued Climax, Gilbert and Battles for misrepresentation, obtaining a patent of his own (US No. 488484 of 20 December 1892) restoring the two-speed gearbox. About fifty locomotives of the perfected Scott design were then made by the Dunkirk Engineering Company.

Designed by Charles Heisler of Dunkirk, New York, and protected by US Patent No. 482828 of 20 September 1892, Heisler locomotives were made until 1941 first by the Stearns Manufacturing Company and then by the Heisler Locomotive Works of Erie, Pennsylvania. A two-cylinder 'V'-form engine drove a longitudinal shaft on the centre line.

The Scene in 1914

Édouard Sauvage once suggested that 1904–13 had been 'a period of fourteen years [in which] loads have been doubled but coal consumption remains much the same', a testament to the greatly improved performance of the steam locomotive. In the same decade, French railways had acquired compound locomotives in great numbers: of 1,363 engines purchased in 4-4-2, 4-6-0, 4-6-2 and 4-6-4 notation, 1,246 had been four-cylinder compounds. The largest individual purchaser, the PLM, had taken 527 locomotives – twenty Atlantics (4-4-2), 330 Ten-Wheelers (4-6-0) and 177 Pacifics (4-6-2), 436 being compounds and 91 simple expansion. In Britain, only a single Pacific, Churchward's *The Great Bear*, had been made when the First World War began. Many British passenger trains were still being hauled by simple expansion 4-4-0s.

Sauvage also reported a trial made with two PLM Pacifics, simple 6102 and compound 6204, on runs between Laroche and Dijon. The compound took a heavier train – 646 tonnes behind the tender compared with 487 – and developed more power, indicator diagrams giving a maximum *cheval vapeur* of 2,425 compared with 2,051. Use of coal also favoured the compound, which, despite the greater train load, burned an average of 1.12kg/cv/km compared with 1.68kg for the simple expansion locomotive; use of water was reduced by much the same ratio. In French eyes, therefore, the compound could do no wrong.

Other European countries were not so certain. Notable advances were made in Britain by George Jackson Churchward (1857–1933), who became chief mechanical engineer of the Great Western Railway in 1900, succeeding William Dean, and immediately implemented many lessons he had learned from trips to North America. The GWR possessed a wide variety of obsolescent locomotives, often with outside frames and external cranks, but, within a decade, Churchward had introduced a range of simple expansion designs – including 4-4-0, 2-6-0, 2-6-2T, 4-6-0 and 2-8-0 – based on a handful of standard boiler types. He was not always given due credit, but the GWR motive power roster in 1948, when British Railways took over, would not have been unfamiliar. Many of Churchward's locomotives were still working effectively.

The largest component of the railway network in the German Empire, the *Königlich Preussische und Grossherzoglich Hessische Staatseisenbahnen* or Royal Prussian and Grand-ducal Hessian state railways, had a few compounds of its own. But these were often pre-1900 von Borries designs that found very little favour with the regime of Robert Garbe.

Garbe (1847–1932) was a highly talented engineer, but his mantra was simplicity. Consequently, the Prussian ethos was one of reliable, trouble-free service in which speed counted for comparatively little. The advent of the Schmidt superheater, which made much the same improvements to saturated steam simple-expansion locomotives as compounding could do, played into the administration's hands. That compounding and superheating could make an even greater difference when used together went unacknowledged.

A four-cylinder superheated simple expansion express passenger locomotive appeared in 1910, but though more than 200 S10s had been purchased by 1914, performance was not good enough; the four-cylinder de Glehn-type S10[1] compound developed by Henschel in 1911 proved to be far better.

The quest for simplicity at the expense of performance was not favoured elsewhere in Germany. In Bavaria, for example, the *Königliche Bayerische Staats-Eisenbahnen* faced different challenges. Parts of the railway network were surprisingly hilly, and so the majority of the passenger locomotives ordered after 1895 had been four-cylinder compounds. In 1899 and 1900, the railway authority imported four Vauclain compounds made by Baldwin in the USA to compare with existing motive power. The Baldwin 2-8-0 and 4-4-2 proved to be efficient enough to influence Bavarian practice, though the superposed cylinders of the Vauclain system of compounding were replaced with an adaptation of the von Borries type.

Bavarian locomotives were purchased from privately owned manufacturers, and so much of the detailed design work on the locomotives to be derived from the Baldwins was delegated to Maffei.

J. A. MAFFEI, MÜNCHEN 2

$^{3}/_{6}$ (2-C-1) gek. Vier-Zylinder-Verbund-Schnellzuglokomotive „S $^{3}/_{6}$" der Königlich Bayerischen Staatseisenbahn, gebaut von J. A. Maffei, München.

Durchmesser der Hochdruckzylinder	425 mm	Triebrad-Durchmesser	2000 mm
Durchmesser der Niederdruckzylinder	650 mm	Heizfläche des Kessels	268 m²
Kolbenhub	670 mm	Dienstgewicht	88 t
		Spurweite	1435 mm

The Bavarian S3/6 Pacific, from a picture postcard published *c.* 1913.

Anton Hammel (1857–1925), Maffei's chief engineer and principal designer, was responsible for the S 2/6 of 1906, a large-wheeled 4-4-4 passenger locomotive built for the Nürnberg Exhibition, which briefly held the official world speed record for railway traction: 154.5kph (96mph), attained in July 1907.

The S 2/6 had four cylinders in line across the bar frames, two low-pressure internally and two high-pressure externally; divided drive; and streamlined fairings on the buffer beam, the chimney, the steam dome and the cab front. Tests soon showed that the locomotive had considerable potential, but that predicted growth in traffic and train loads would require something more powerful. Hammel's answer was the S 3/6 of 1908, an elegant four-cylinder compound 4-6-2 derived from an engine that had been supplied to the railway of the grand duchy of Baden. The S 3/6 proved to be outstandingly successful, smooth running and easily capable of 120kph (75mph) with a substantial train. Indeed, when the German state railways were nationalised in the early 1920s, seventy additional S 3/6 were ordered to plug the most obvious gap in the roster of ex-Prussian classes: the absence of anything powerful enough to pull an express train of any note!

There were two basic types of S 3/6, as a few locomotives were made in 1912 with driving wheels of 2m (6ft 7in) instead of 1.87m (6ft 2in). They were destined for the comparatively flat main lines between München and Nürnberg or Würzburg, where the highest speeds were permissible. The modernised Deutsches Reichsbahn Gesellschaft (DRG) 'Klasse 18.4' lasted in service long after the Second World War had ended; the last survivor was not withdrawn until 1969.

3

SUPREMACY

Steam at the Height of its Power, 1919–55

In the aftermath of the First World War, railways stood at a crossroads. Though the devastation had been confined largely to the Western Front, where many communities in southern Belgium and northern France had been ground to dust, many lessons had been learned by operating railways in adverse conditions. Both sides had made large numbers of narrow-gauge steam locomotives to supply the armies, but the vulnerability of boilers to shell fragments had allowed internal combustion engines to mount the first serious threat to steam's supremacy.

The values of standardisation, already well established in Prussia, were increasingly clear. Even the British had sought a large number of standard-gauge heavy freight locomotives to serve in France, seizing on the Robinson 2-8-0 of the Great Central Railway as an appropriate pattern. These locomotives, now often known as 'ROD' (Railway Operating Division), were essentially simple and largely trouble-free, with two outside cylinders and a well-tried boiler. The Germans had built substantial numbers of 0-8-0 tank engines for the narrow-gauge supply lines, fitted with Klein-Lindner axles to negotiate sharp curves, and Baldwin had supplied the Allies with many 4-6-0 tank engines for much the same type of service behind the Western Front.

Peace brought new problems. Railways had been required to move men and munitions on an unprecedented scale, but this had shown that many locomotives were too small for the tasks they had been set. There were too many individual designs to allow effectual maintenance programmes to be developed; very few parts interchanged, creating widespread unavailability of motive power; and much of the rolling stock was in a very bad way by 1919. Once powerful railway companies, which had paid their shareholders worthwhile dividends prior to 1914, were no longer financially sound (or, in the case of the French Chemins de Fer du Nord, wrecked by war).

In Britain and in Germany, an answer was found by amalgamating or 'grouping' railways geographically, large and small, to improve their effectiveness. In Britain, from 1 January 1923, this created the London, Midland & Scottish Railway (LMS), the London & North Eastern Railway (LNER) and the Southern Railway (SR), though the Great Western retained something of its independence. In Germany, the *Deutsches Reichsbahn Gesellschaft* (DRG) was created from the previously independent state railways.

In North America, where the larger railroad companies had gradually enveloped many of the 'Short Lines', the US Railroad Administration (founded on 28 December 1917, two days after President Wilson had ordered that all railroads should be placed under national control for the duration of hostilities) had promoted a series of standardised locomotives

The Great Central Railway 2-8-0 of 1911, designed by John G. Robinson, had 4ft 8in driving wheels, two 21 × 26in cylinders and a boiler pressure of 160psi (later raised to 180psi); locomotive weight was 73 tons 1 cwt. More than 500 were purchased during the First World War by the Railway Operating Department (ROD). These had Westinghouse air brakes on the smokebox and train-heating capability. Many were bought by individual railways, including the GCR and the GWR, once hostilities had ended.

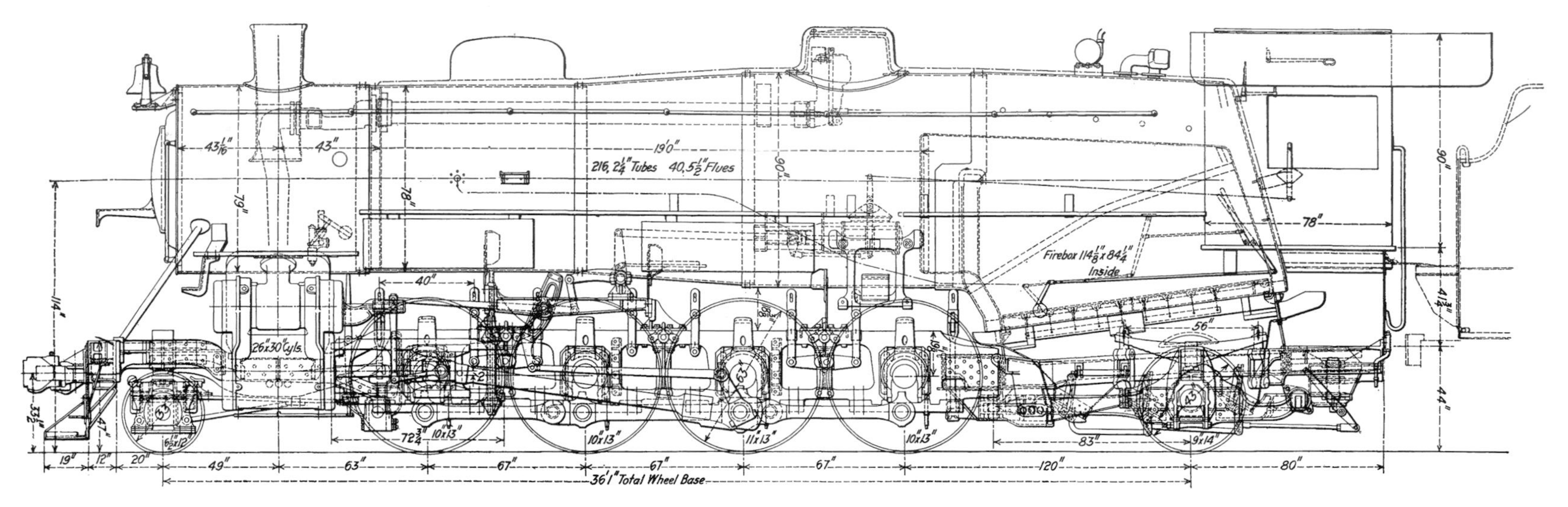

A typical USRA standard design: the light 2-8-2 Mikado was among the most numerous of the USRA standard designs: more than 600 were made in the 1920s. (*The U.S. Railroad Yearbook*, 1927)

ranging from the smallest 0-6-0 Switcher to a heavyweight 2-8=8-2 Mallet, to be made by the leading manufacturers – Alco, Baldwin, Lima – to alleviate a critical shortage of modern motive power. Standardising components, it was hoped, would enable locomotives to be supplied almost to order for the duration of the war. A total of 1,856 had been made when the USRA mandate was withdrawn on 1 March 1920, but work continued until the ultimate total was about 3,250. The most commonly encountered designs were the light 2-8-2 (625 examples), the 0-6-0 (255), the heavy 2-8-2 (233), the 0-8-0 and the heavy 2-10-2 (175 apiece) and the 2-8=8-2 Mallet (106). A synthesis of the best existing railroad practice, they were sturdy and reliable. Features such as power-operated firebox doors, power reverse and mechanical stokers were a revelation to engine crews on many of the lesser railroads.

The same process of standardisation could be seen in Britain, where a programme of widespread scrapping of obsolete and non-standard designs began – much to the horror of lesser companies such as the

Highland Railway, the fate of whose rolling stock was largely in the hands of officials of the hated Caledonian Railway. New locomotives appeared from the principal manufacturers, but their diversity was limited by the appointment of chief mechanical engineers in each of the new groups. By 1930, therefore, the influence of the new regimes was clear to see.

An Indian Locomotive Standards Committee was formed in 1924 to bring order to the proliferation of classes used in the sub-continent on narrow, metre, standard and broad gauges. There was an increasing feeling among Indian railway engineers that locomotive design had been hamstrung by political ties with Britain, and, therefore, that India had got what the manufacturers would supply instead of what was best for the job. This view was undoubtedly exaggerated, yet the locomotives developed by the committee – the XA light, XB intermediate and XC heavy broad-gauge 4-6-2s, the XD light and XE heavy 2-8-2s – were still clearly products of the British railway industry. They were simple, with two outside cylinders and a comparatively modest boiler pressure of 180psi, but experience showed the design of the bogies and trailing trucks to be unsuited to track that was often in very poor condition. When the later WG broad- and YP metre-gauge 2-8-2s were adopted after the end of the Second World War, India (and Pakistan) had gained independence and the effects of the WP passenger locomotives supplied from the USA during the conflict could be seen clearly.

The major independent railways in France were not combined in SNCF until 1938, and so the influence of their designers did not wane appreciably in the 1920s. Standardisation had already been taken to extremes by the Prussians, but the lack of capable express passenger locomotives on the Prussian system forced the DRG not only to make good use of the Pacifics of Bavaria and Württemberg, but also put the Bavarian Maffei pattern back into production as Klasse 18.4. New Pacifics and even a Mikado (based on a Prussian prototype) were introduced as soon as manufacturing capacity could be spared.

For the first time, the supremacy of the conventional steam locomotive was increasingly challenged. The Heilmann steam-electric engine had been tested in the 1890s, but, despite promise, had been regarded as too heavy and too complicated to mass produce. The Stumpf uniflow engine had been tried in Germany and Britain with no lasting results; and the unlucky eight-cylinder Paget locomotive of 1908 had never been given a chance to overcome what were admittedly potentially serious teething troubles. However, the use of turbo-electric drive in warships as large as the US Navy battleship *New Mexico* inspired fresh approaches to be taken.

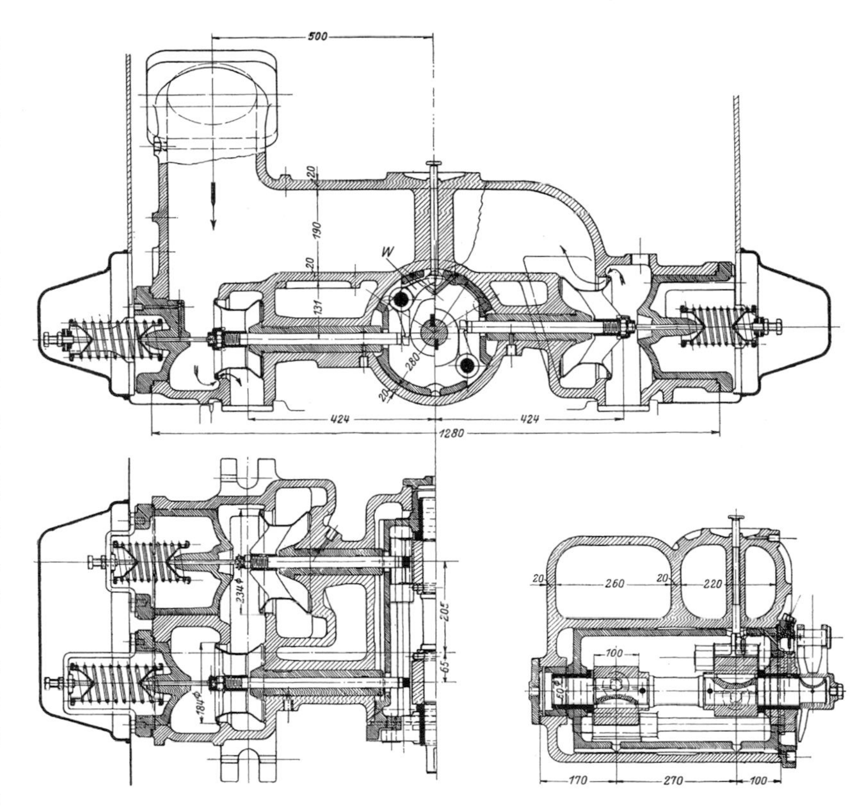

Lentz poppet valve gear was complicated and delicate, but worked well when properly constructed and well maintained. This shows the valve and cam box fitted to 2-8-4 locomotives of the Austrian federal railways. (*The Locomotive*, 15 November 1935)

At their most basic, these were attempts to improve steam flow and develop valves that would give better control over the steam cycle. Among the most noteworthy were the German Lentz and the Italian Caprotti designs, which, in the latter case, were still being tried at the end of the steam era. It could make a difference. One Pennsylvania Railroad K4 Pacific fitted with Franklin oscillating-cam poppet valves installed by Lima Locomotive Works is said to have increased drawbar horsepower by 44 per cent at 80mph. This was enough to persuade the railway management to specify cam gear for the T1 duplex instead of Walschaerts' type.

TURBO-ELECTRIC CONDENSING LOCOMOTIVE.

GAUGE 4′ 8·5″ FACTOR OF ADHESION 11·05′.

TRACTIVE FORCE 22,000 LBS.

WORKING PRESSURE–200 LBS. PER SQ. IN.

WHEEL DIAMETERS	COUPLED	4′ 0″
	BOGIE	3′ 1″
WHEEL BASE	RIGID	16′ 4″
	TOTAL	59′ 4″
		Tons
WEIGHTS	BOILER VEHICLE	67·25
	CONDENSER VEHICLE	63·5
	TOTAL	130·75
	ADHESIVE WEIGHT	108·5
HEATING SURFACE WITH SUPERHEATER		1453 sq. ft.
GRATE AREA		28.4 „ „
TURBO GENERATOR		890 K.W. at 3,600 revs. 27·5″ vacuum.
TANK WATER CAPACITY		2500 Galls.
FUEL CAPACITY		4 Tons
VOLTAGE		600 volts.

REF. No. E 14.

LOCOMOTORA DE CONDENSACIÓN TURBO-ELÉCTRICA.

ANCHO DE VÍA: 1,43 METROS. FACTOR DE ADHERENCIA: 11,05.

ESFUERZO DE TRACCIÓN, 9.979 KGS.

PRESIÓN ÚTIL: 14,061 KGS. POR CM. CUADRADO.

DIÁMETROS DE LAS RUEDAS	ACOPLADAS	1,22 m.
	BOGIE	0,94 m.
DISTANCIA ENTRE LOS EJES	RÍGIDAS	4,98 m.
	TOTAL	18,08 m.
PESOS	VEHÍCULO DE CALDERA	68,32 ton.
	„ „ CONDENSADOR	64,51 „
	TOTAL	132,84 „
	ADHERENCIA	110,23 „
SUPERFICIE DE CALDEO CON RECALENTADOR		134,98 m²
SUPERFICIE DEL EMPARRILLADO		2,64 „
TURBO-GENERADOR		890 K.V. a 3.600 r.p.m. Vacío de 698,50 mm.
CAPACIDAD DEL DEPÓSITO DE AGUA		11365 lit.
CAPACIDAD DE COMBUSTIBLE		4,06 ton.
VOLTAJE		600 voltios.

No. DE REF. E 14.

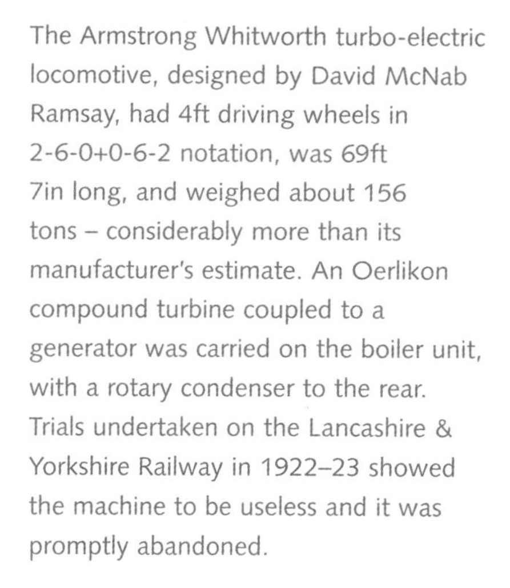

The Armstrong Whitworth turbo-electric locomotive, designed by David McNab Ramsay, had 4ft driving wheels in 2-6-0+0-6-2 notation, was 69ft 7in long, and weighed about 156 tons – considerably more than its manufacturer's estimate. An Oerlikon compound turbine coupled to a generator was carried on the boiler unit, with a rotary condenser to the rear. Trials undertaken on the Lancashire & Yorkshire Railway in 1922–23 showed the machine to be useless and it was promptly abandoned.

The most promising of the radical solutions seemed to lie in the turbine. The Reid-Ramsay Electro-Turbo Locomotive was tested in Britain in 1910, but a development by Armstrong Whitworth, in accordance with British Patent 163538 granted to David McNab Ramsay and James William Wood of Glasgow on 26 May 1921, proved to be no better when tested by the Lancashire & Yorkshire Railway in 1922–23. Ramsay, who had trained as an engineer in the Caledonian Railway works in St Rollox, strove for some years to improve his locomotive, but the Zoelly and comparable designs tried in Europe promised much better results.

The Swedish Grangesberg–Oxelosund railway (TGOJ) purchased three M3t 2-8-0 locomotives, built by Nydqvist & Holm with Ljungstrom non-condensing turbines and jackshaft drive. Introduced to traffic in 1937, they performed reliably until withdrawn in the mid-1950s. The London Midland & Scottish Railway built one of the Princess Coronation class as the Turbomotive, No. 6202, which ran successfully until the turbine failed in 1949.

In the USA, some steam turbo-electric locomotives of prodigious dimensions were introduced to the Chesapeake & Ohio and Northern &

The Gresley Hush-Hush of 1929 was a four-cylinder compound with a Yarrow-Gresley high-pressure boiler (450lb/in^2). The wheel notation is customarily listed as '4-6-4' (Baltic), but the trailing axles were separated and '4-6-2-2' was often preferred at the time No. 10000 was running on the London & North Eastern Railway. The boiler proved to be unsatisfactory, so the locomotive was altered in 1936 to three-cylinder simple expansion.

Western railroads, but utterly failed in service. By far the most impressive was the 6-8-6 non-condensing turbine locomotive built by the Pennsylvania Railroad in 1944, which combined a conventional steam locomotive layout with great power. This locomotive demonstrated on several occasions that it could do what the Pennsylvania T1 4-4÷4-4 duplex could not: pull a 1,000-ton train at 100mph!

Another false dawn was the high-pressure boiler. Water tube designs such as the Brotan (particularly popular in Hungary) were made in surprising numbers; however, the rigours of service, and the dangers of pressures in excess of 900psi, inhibited progress. The experimental British 4-6-0 *Fury* failed, with loss of life, and was subsequently reintroduced as a member of the Royal Scot Class. The Gresley 'Hush-Hush' 4-6-2-2 of 1929, which had a Yarrow high-pressure boiler, also failed to convince its doubters and was duly converted to conventional form in 1936.

High-pressure boilers were used successfully at sea, but the stresses and strains created when a locomotive ran over uneven track usually proved to be too much for the integrity of joints in pipework. Unfortunately, turbines run most efficiently at full speed and are wasteful of steam (and, therefore, fuel) at low power. Nor can they be reversed. Ships were often fitted with separate cruising and reverse turbines, but space was not generally at a premium. Though the Swedish iron ore locomotives were non-condensing, it is generally agreed that condensers are essential if turbines are to run efficiently. Unfortunately, their size, complexity and fragility brought penalties severe enough for most railways to lose interest.

While engineers struggled to improve the steam locomotive, the advent of high-speed diesel railcars such as the *Fliegender Hamburger* seized public attention and highlighted a desire for speed. However, though the railcars were glamorous, their haulage capacity was still comparatively limited; the diesel locomotive had yet to outperform its steam rivals, the best of which could haul far greater loads at generally comparable speeds. The quest for speed not only produced some of the most impressive of all railway engines but also contributed greatly to thermodynamic improvement.

Attempts had been made in France as early as 1882–83 to reduce the resistance offered by the customarily bluff-fronted locomotive to the wind. This is generally now associated with the Chemins de Fer PLM, whose *Coupe-Vents* or 'Windcutters' were intended principally for the line that ran to Marseille and the south of France by way of Lyon. Fairings on the smokebox, chimney, cab and sometimes over the cylinders had been shown to reduce the drag when the train was headed directly into the Mistral that blew down the Rhône valley at speeds of 60mph or more.

Except for a few cab-forward types developed in France, Germany and Italy prior to 1914, few dedicated attempts to streamline locomotives were made elsewhere until the belated acceleration of steam-hauled services in the 1930s. Then there were several catalysts: the need to haul larger trains in circumstances where the power of locomotives was limited by restrictive dimensions or geography; attempts to restore the image or use of the railways in a period of economic depression (especially in the USA); and the growing awareness of the public relations value of industrial design.

Boston & Maine P-4 class 4-6-2 No. 3710 was built by Lima Locomotive Works in 1935. Note the 'air smoothed' fairing on the boiler. The locomotive had 6ft 8in driving wheels, two 23 x 28in cylinders and a boiler pressure of 260psi; a booster was originally fitted to the tender, but was soon abandoned.

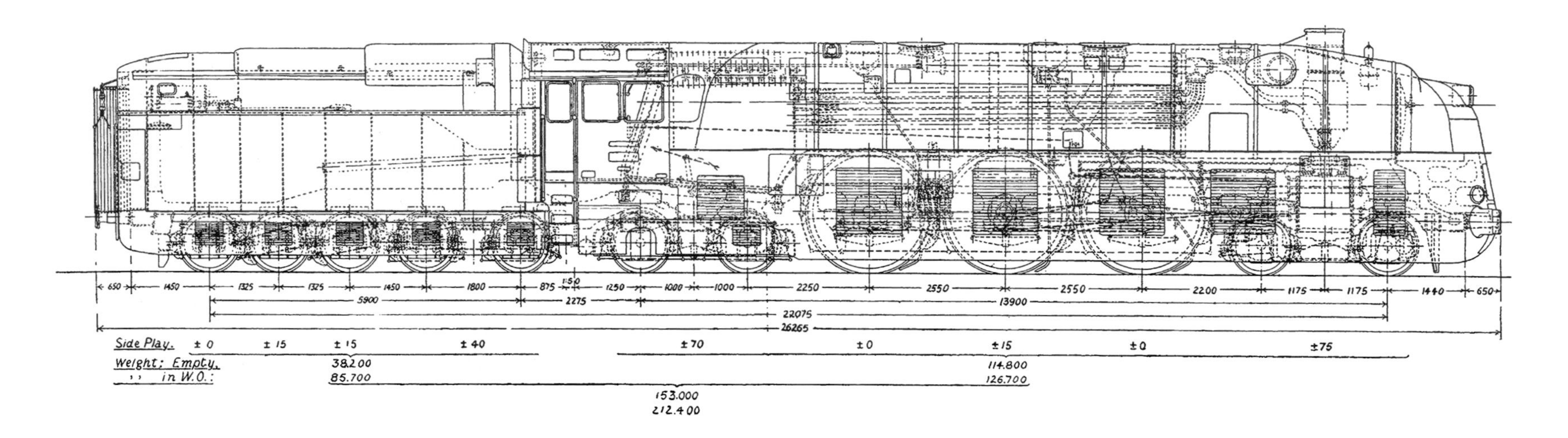

The German streamlined 05 Class 4-6-4 was built by Borsig Lokomotiv-Werke to pull 250-tonne trains at 175kph. Three of these three-cylinder simple expansion locomotives were made, with 2.28m (7ft 6in) driving wheels and a boiler pressure of 19.5at (287psi). One of them, 05 003, even had a cab-forward layout and an unsuccessful pulverised-coal stoking system. (*The Locomotive*, 15 April 1935)

There are several plausible claimants for the title 'first to exceed 100mph' (not *City of Truro* or even *Flying Scotsman*!), beginning with a P5-Class Camelback 4-4-2 of the Philadelphia & Reading Railway during a high-speed run on 14 June 1907. But it was obvious that there was a finite speed limit for steam traction, deciding factors including coupled-wheel diameter and piston speed. Though Gresley Pacific *Papyrus* attained 108mph during trials of 5 March 1935, streamlining was deemed to be advantageous. The die was cast when brand-new 4-6-4 streamliner 05 001 claimed the record for Germany on 11 May 1936, attaining 124.5mph. This was bettered by the LNER 4-6-2 *Mallard*, which sustained 125mph with a short duration burst of 126mph down Stoke Bank in the summer of 1938.

These figures were almost always attained with lightweight trains, but they did point the way to better things; both the LNER and the LMS ran streamlined services that were capable of cruising at more than 100mph for long distances. Commercial rivalry in the USA produced not only an extraordinary series of streamlined locomotives but also some impressive 'speed with load' records.

Gresley A4 class No. 4496 *Golden Shuttle*, outshopped from Doncaster in September 1937, was renamed *Dwight D. Eisenhower* in September 1945 and withdrawn in July 1963 as British Railways No. 60008. The fairings over the driving wheels were removed in 1942 to facilitate maintenance in wartime conditions.

There were several approaches to streamlining. In Britain, France and to a lesser extent in Germany, Belgium and the Netherlands, the goals were aerodynamic efficiency and methods of keeping smoke away from the cab. Shrouding varied from a few fairings – Collett's attempts on a Castle- and a King-Class 4-6-0 for the Great Western Railway were almost risible – to the almost total enclosure of the German 05 Class 4-6-4, which even included roller shutters in the panels that gave access to the motion.

The A4 streamlining had been developed in collaboration with Professor W.R. Dalby. Models had been tested in a wind tunnel and it had taken a surprising amount of time to find the contours that would lift smoke clear of the cab. Experience with the A4 locomotives soon showed that, compared with A3, considerably less fuel was burned to reach a particular speed. But the factor of improvement was difficult to quantify.

In the USA, public relations had the most dynamic influence. Beginning with *Commodore Vanderbilt* and the inauguration of the *20th Century Limited* in February 1935, and ending with *City of Memphis* in 1947, more than 200 locomotives and matching trains took to the rails in North America. The principal railroads employed industrial designers to shroud the machinery: German-born Otto Kuhler (1894–1977), Henry Dreyfuss (1904–72) and Frenchman Raymond Loewy (1893–1986) are perhaps the best known. Kuhler styled the *Hiawatha* of the Milwaukee Road; Dreyfuss

A poster by Leslie Ragan advertising the 'New 20th Century Limited' run by the New York Central Railroad. The Dreyfuss styling shows that it post-dates 1937, heralding a golden era. Railroads strove to dominate as the USA left the Great Depression behind, and steam was facing the very real challenge of diesel and electric traction.

New York Central J-3a Hudson (4-6-4) No. 5453, built by Alco in 1938, shows the benefits of Henry Dreyfuss' air-smoothed casing. This particular locomotive has Scullin-type wheels instead of the Boxpok type.

was responsible for the *20th Century Limited* of the New York Central; and Loewy designed the shrouding of the Pennsylvania Railroad locomotives.

Some railroads simply applied shrouds of varying degrees of sophistication to existing locomotives, generally of Pacific (4-6-2), Hudson (4-6-4) or Northern (4-8-4) notation. The oldest were three 1910 vintage Pacifics converted in December 1939 to pull *The Firefly* of the St Louis & San Francisco Railroad ('Frisco Lines'). These locomotives had proved to be fast and steady runners. Shrouding was inexpensive and also added substantially to the locomotive's weight, improving adhesion.

Livery varied from the black, bright red and silver stripes of the Lehigh Valley's *Black Diamond* to bright green, silver and yellow on the Southern Railway *Tennessean*. The 4-6-4 *Aeolus* locomotives – two, identical – of the Chicago, Burlington & Quincy Railroad were shrouded entirely in stainless steel.

The ability to pull heavy loads at high speed put the giant North American locomotives at the forefront of progress even though they were usually two-cylinder simple expansion designs. *Aeolus* was claimed to have reached 120mph and the 4-6-2 F7-Class 4-6-4 of the Milwaukee Road *Hiawatha* is said to have been capable of 125mph with an eight-coach train. Most extraordinary of all were the fifty Pennsylvania's 4-4÷4-4 T1 duplex locomotives, one of railway history's fiascos, which were intended to haul 1,000-ton trains at more than 100mph. Though prone to slip violently if too much power was applied at the wrong moment, there is little doubt that the T1 was very fast; rumours still persist that 125mph could be exceeded – much to the chagrin of Gresley A4 supporters! – and an enthusiasts' group is currently raising funds to enable a new T1 to be constructed.

By the 1930s, efforts were also being made to improve the thermal efficiency of the steam locomotive, much of the best work being undertaken in France. Every major pre-nationalisation French railway used

A French Pacific of the Nord railway, No. 3 1249, with Chapelon improvements and Cossart valve gear.

compounds, four-cylinder patterns being most common, though a few had two cylinders (usually rebuilds of old stock) and the post-1938 SNCF designs almost always had three. The encouragement of compounding in the period between the world wars was largely due to the genius of André Chapelon, whose 1929 transformation of a Paris–Orléans 4-6-2 had verged on miraculous. From an engine that could develop about 1,850ihp, Chapelon obtained 3,000ihp at 75mph. Later conversions raised this to 3,700ihp, largely owing to streamlined steam passages and oscillating-cam poppet valves driven by Walschaerts' gear. A similar transformation was made of the Paris–Orléans 4-8-0.

André Chapelon, born in 1892, served in the artillery during the First World War and joined the Paris–Lyon–Mediterranée Railway in 1919 before moving in 1925 to the Chemins de fer Paris–Orléans. There he became principal experimental engineer in 1935. His achievements are legendary: doubling the tractive effort of some locomotives by careful modification and an astute appreciation of the underlying thermodynamic theory. The first 'Chapelonised' 4-6-2 appeared in 1929, followed in 1932 by the 4-8-0. When the French national railway (SNCF) was formed in 1938 by grouping together the principal independent lines, Chapelon was employed by the department of steam locomotive studies (from which he retired in 1953 as head). Progress with his designs was interrupted by the Second World War, but trials undertaken in the late 1940s with 4-8-4 and 2-12-0 designs eclipsed virtually everything that had gone before; Chapelon's locomotives generated much the same power as North American Malletts, which weighed more than twice as much. However, there were many who had vested interests in undermining Chapelon's research – those, for example, who were determined to introduce electric traction at virtually any cost. A campaign of whispered criticism, misleadingly skewed statistics and the peremptory scrapping of the epitome of steam locomotive designs says it all. Chapelon subsequently worked as a consultant and died in 1978.

Chapelon's masterpiece was 4-8-4 No. 242A1, which could generate 5,500ihp in the cylinders (equivalent to about 4,200hp at the drawbar). However, the sophisticated analytical approach of Chapelon and his champions found very little favour in North America, where, for many, size still mattered. Improvements in manufacturing techniques allowed simple expansion versions to be made with boilers pressed to new heights, even though the two 'Standard Mallets' promoted by the US Railroad Administration in the immediate post-1919 period had been compounds.

The principal problems were keeping flexible steam joints tight and ensuring that pressure or temperature did not drop as steam traversed the length of the pipework connecting boiler and cylinders. Eventually, the low-pressure cylinders became so large that most railways took the decision to build superheated simple expansion Mallets such as the Union Pacific 4-6=6-4 Challengers and 4-8=8-4 Big Boys – even though this required considerable redesign to ensure that steam reached the remotest cylinders at full pressure. Only the Norfolk & Western was still commissioning compound 2-8=8-4 Mallets when the last steam engines were built in the USA in the 1950s.

The first simple expansion Mallet (not what the inventor had intended!) seems to have been a 2-4=4-0 in use on Trans-Siberian Railway in 1902, followed by a 2-8=8-2 and a 2-8=8-0 on the Pennsylvania Railroad (1912 and 1919 respectively), but none were notably successful. A simple expansion 2-6=6-2 was made by Baldwin for the Baltimore & Ohio Railroad in 1930, with 70in-diameter driving wheels, and a 2-6=6-4 was supplied to Seaboard Air Line in 1935. But the first truly successful simple expansion Mallet was the 4-6=6-4 *Challenger* introduced by the Union Pacific Railroad in 1936.

The twenty-five examples of the Union Pacific Railroad Big Boy-Class 4-8=8-4 Mallets of 1941–44 were the largest steam locomotives ever built. Made by the American Locomotive Company, Big Boys had four enormous cylinders – 23¾ × 32in – and 5ft 8in driving wheels. A boiler pressure of 300psi gave a nominal tractive effort of 135,375lb. The locomotives could develop more than 6,000hp at the drawbar, and could haul 100-car freight trains weighing up to 4,000 tons over a hilly section of the Union Pacific. Though speeds rarely exceeded 40mph, the huge machines could reach 80mph if pushed. A typical example of the 1941 batch measured 132ft 10in overall, 16ft 2½in high, 10ft 10in wide, and weighed a staggering 595 short tons. The enormous tenders held 24,000 US gallons of water and 28 short tons of coal.

One of the Union Pacific's 4-8=8-4 'Big Boy' Mallets, No. 4007 (Alco, 1941), 'on the road' near Laramie, Wyoming. (Dr Richard Leonard; photograph taken by David Leonard in August 1957)

Whereas the Mallet was at its best in the USA, all but unencumbered by loading-gauge restrictions and assured of good-quality track, the Garratt was favoured in areas where heavy loads had to be hauled over undulating, lightweight track with severe curves. The competing designs rarely co-existed: the Garratt ousted its rival from southern Africa, whereas the success of the Mallet in North America prevented the Garratt attaining even a toehold. Only in South America (particularly Brazil) and in India were there Garratts and Mallets in quantity.

It is a shame that no one ever tested a Garratt built to North American loading gauge in the USA, where it undoubtedly had advantages over its rival; the boiler could be made much more compact, and the separation of the drive units, both of which were pivoted, allowed sharper curves to be followed. The largest of the Garratts used in southern Africa showed what could be done even on comparatively lightweight track, where a Mallet of comparable power would not have been able to venture.

Patented in 1907 by Herbert W. Garratt, inspecting engineer of the New South Wales Government Railways, the principal claim to novelty was the attachment of a short, large-diameter boiler unit to the end of two power trucks. Each truck carried its own water tank or bunker, helping to restrict the strain on the pivots by relieving the boiler assembly of unnecessary weight. The open space beneath the main frames, unencumbered by axles, wheels and valve gear, allowed a deep fire grate and a readily accessible ash pan to be fitted. Together with the excellent proportions of the boiler, these features gave excellent steam-raising capacity. In this respect the Garratt was more effectual than the Mallet.

The earliest Garratt was built in 1909 for 2ft-gauge North East Dundas Tramway in Tasmania by Beyer, Peacock & Co. Ltd. It was also a compound, a type of operation rarely favoured for Garratts – few of which used anything other than four- or occasionally six-cylinder simple expansion. Virtually all Garratts were built in Manchester, except for a few made in France and Belgium for service in Algeria and a handful built under licence in Spain. The largest was a solitary 4-8-4+4-8-4 Class 10 Ya, built for the USSR in 1932, which weighed 262.5 tons and had a nominal tractive effort of 78,700lb.

The largest steam engine to run in Britain was LNER 2-8-0+0-8-2 Garratt Class U1 No. 2325, built in 1925 by Beyer, Peacock & Co. Ltd in Manchester. The solitary example was intended to bank 1,100-ton coal trains on the incline from Wath to Penistone. Weighing 178 tons in working order, it had six 18½ × 26in cylinders which, with a boiler pressure of 180psi and driving wheels 4ft 8in in diameter, gave a nominal tractive effort of 72,940lb. No. 2325 was withdrawn for scrapping in the mid-1950s.

Beyer, Peacock & Co. Ltd gave consideration to a 'Super Garratt', conceived by Robert Harben Whitelegg (once locomotive superintendent

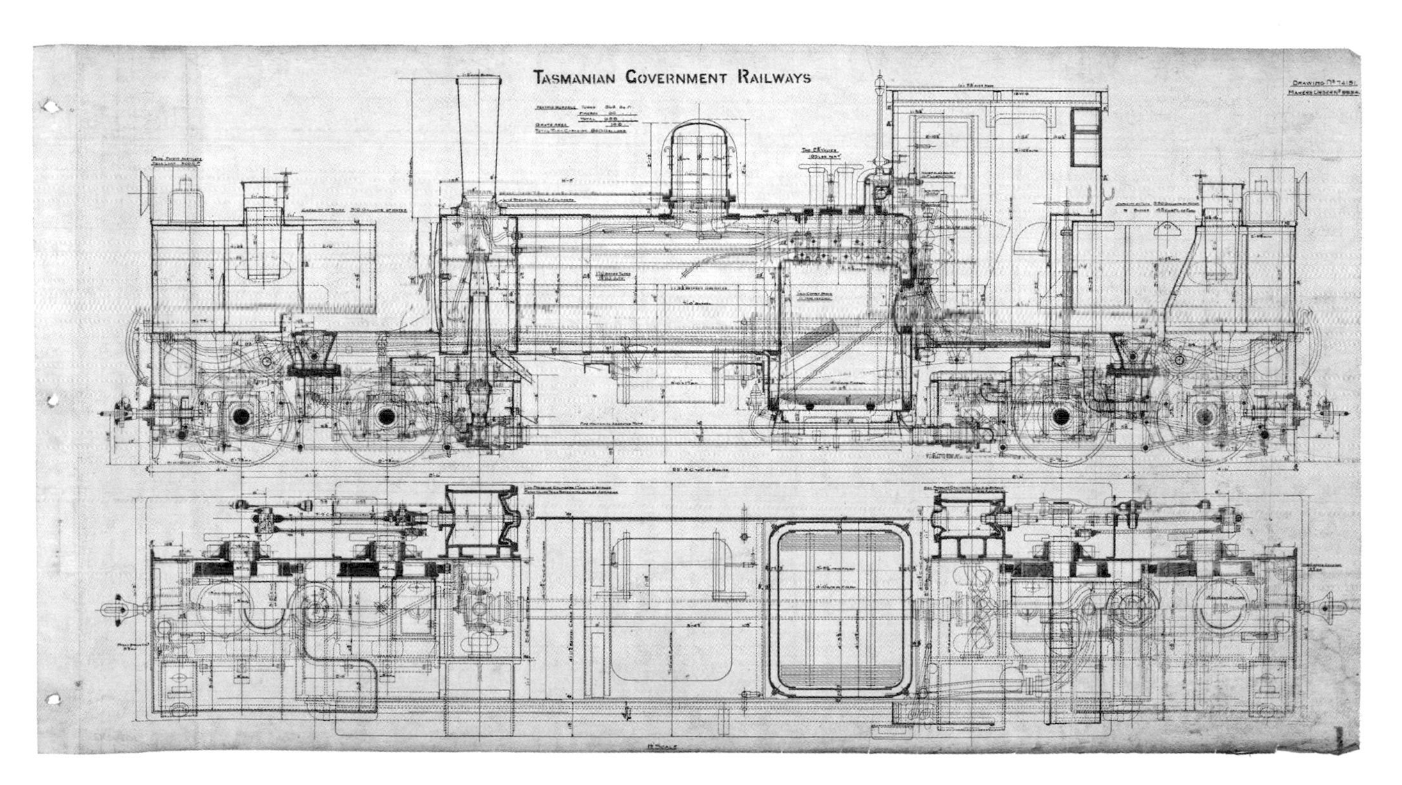

Beyer, Peacock & Co. Ltd drawing No. 74151 shows the first Garratts to be made, a pair of Class K 0-4=4-0s destined for Tasmania. (National Archives of Australia, NAA P1297,1)

of the London, Tilbury & Southend and Glasgow & South Western railways). British Patent 230511 of 11 March 1925 shows a turbine derivative of 0-8=8-0+0-8=8-0 notation, with the fore bogies pivoted in pseudo-Mallet fashion beneath the power unit; British Patent 230888 of 25 March 1925 protected a more conventional 0-6=6-0+0-6=6-0 version, apparently simple expansion. Neither Super-Garratt was built, but a 'Union Garratt' was made in small quantities in Germany by Krauss-Maffei. This combined some Garratt and Mallet features. Though the independent power units were retained, the boiler and the bunker were carried on a rigid frame running back over the rear unit. This allowed a boiler of more conventional form to be used.

When the Second World War began in Europe in 1939, railways once again showed their value. They allowed the Germans to move men and matériel first eastward into Poland, then westward into Belgium and France, and eastward again into the Soviet Union. But railways also allowed the Russians, with superhuman effort (and a cost in lives that should never be undervalued), to move large parts of vital industry out of German hands. This feat alone ultimately won the war, as it allowed the Red Army to outgun the invaders. Railways also played a critical part in the build-up to D-Day.

The ferocity of the Second World War brought the destruction of rolling stock on an unprecedented scale. Steam locomotives were vulnerable to a single piece of shrapnel, if it struck the boiler, and most combatants realised that new locomotives were almost as vital to the war effort as tanks. The most impressive programme produced the German *Kriegslokomotive*, several classes of which were still being delivered in huge numbers when the war ended.

The earliest examples were comparatively conventional, but design elements were greatly simplified as the industrial situation deteriorated. The Model 50, a conventional design adopted in March 1939, was supplemented by the somewhat simplified Model 50 *Übergangs-Kriegslokomotiv*, but the situation deteriorated until the government intervened in March 1942. The Model 52 embodied as many manufacturing expedients as possible, to save labour and materials, and production began at the end of the year. Eventually, twenty manufacturers and erectors were involved. Shortages of raw material were a particular worry and, particularly after the Allied invasion of Europe, the inability of the German rail network to deliver material exacerbated the problem. Yet total production of the Model 52, in all its forms, amounted to 6,239 (1943–46). The locomotive had 1.4m (4ft 7in) driving wheels, two 60 × 66cm (23⅝ × 26in) cylinders, a boiler pressure of 16at (235psi), and weighed about 84.7 tonnes (bar-frame version) without the tender.

The British had begun simply by making as many locomotives as possible to established designs. The principal beneficiary of this was the Stanier '8F' 2-8-0 of the London, Midland & Scottish Railway, which was accepted for overseas service with only minor modifications. However, in an attempt to conserve material and machine time, a Ministry of Supply team led by Robert Riddles produced first a 2-8-0 and then a 2-10-0 'Austerity' on the basis of the Stanier 8F. The 2-10-0, the first of its notation to run in Britain, proved to be much more successful than predicted and laid the basis for the post-war British Standard 9F design.

Once the Allies had invaded Europe and the war had subsequently ended, there was a great need of motive power on the devastated railway networks. This was solved partly by repairing as many existing locomotives as possible, but all too often the repair facilities had also been razed to the ground.

Another answer was found in the development of simple expansion locomotives, usually based on US practice but adapted to European loading gauges. Engines of this type included 120 'Liberations' designed and built

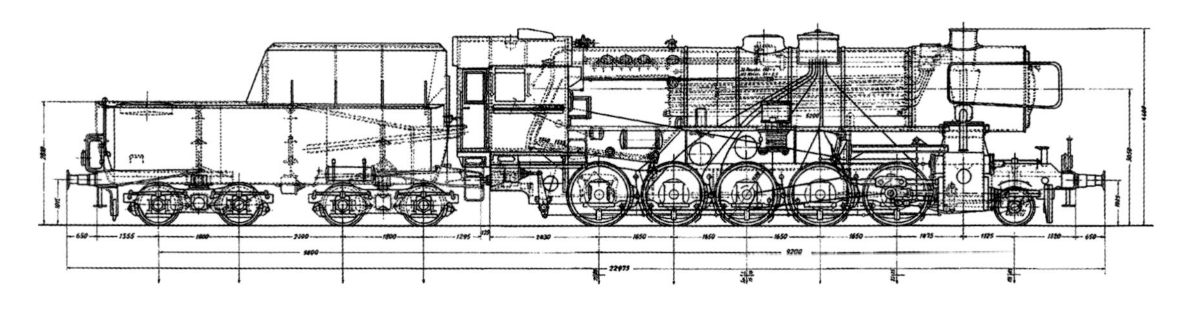

A longitudinal section of a German Modell 52 *Kriegsdampflokomotiv* (KDL-1). More than 7,000 were built from 1942 until the early 1950s, and the last survivor was not withdrawn until 1988. A typical example was 23m (75ft 5in) long, weighed 84 tonnes (engine only), had 1.4m (4ft 7in) driving wheels, two 60 × 66cm (23⅝ × 26in) cylinders and a boiler pressure of about 16at (235psi).

The Liberation-type locomotives made in 1946 by Vulcan Foundry for the UNRRA-incorporated traditional British and North American practices.

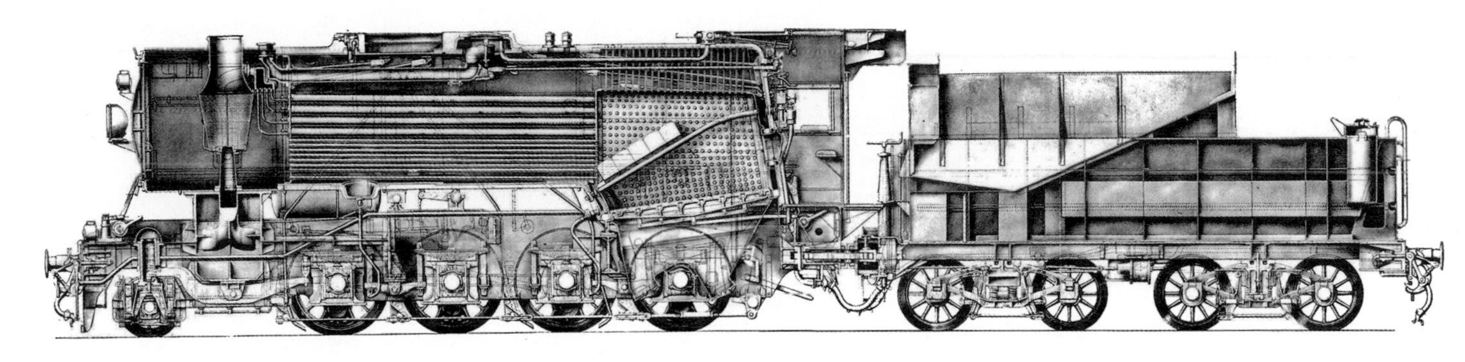

"Liberation" 2-8-0 Type Locomotive and Tender built for U.N.N.R.A. and Luxemburg, 1946

(For use in Poland, Yugoslavia, Czechoslovakia and Luxemburg)

in Britain in 1946 by the Vulcan Foundry of Newton-le-Willows (though very North American in detail), and the USTAC S-160 designed by Major J.W. Marsh of the US Army Corps of Engineers; 2,120 S-160 locomotives were made in 1942–46 by Baldwin, Lima and Alco. Many of these were stockpiled in Britain in the prelude to D-Day and then shipped to Europe, where they were transferred in 1947 to UNRRA and thereafter to national railway systems. Others served in Africa and the Far East.

But then there was Oliver Bulleid, the maverick chief mechanical engineer of the Southern Railway. Somehow, citing lack of motive power, Bulleid persuaded his masters to sanction a 'stripped down' 0-6-0 and two classes of 4-6-2 in the middle of the Second World War. The Merchant Navies and their lighter cousins, the West Country/Battle of Britain type, were listed as 'mixed traffic' when they were clearly destined for passenger services. They were crammed with unusual features, not least being 'air smoothed' casings, chain-driven valve gear sealed in an oil bath and a high-pressure boiler with thermic siphons. Service soon revealed shortcomings; the chains were apt to stretch, altering valve timing, and the locomotives were notoriously heavy on coal. The oil bath leaked on to hot pipework, causing fires, and the casing proved a poor smoke lifter.

Bulleid's outlandish Q1 was the most powerful 0-6-0 ever to run in Britain. Virtually every pretence to aesthetics was abandoned, but savings in weight allowed a capacious boiler to be fitted. However, the absence of a running plate proved to be a mistake: water tended to be thrown up from the wheels in front of the cab windows. BR 33016, formerly C16, was one of twenty built in Brighton Works in 1942; the locomotive was withdrawn in 1963 and scrapped.

The 2-10-0 Class 9 was among the most successful of the British Standard designs. This is 92220 *Evening Star*, outshopped from Swindon in March 1960, the last steam locomotive to be made for British Railways. Withdrawal came in March 1965, after a service life of just five years!

When British Railways was created in 1948, consideration was actually given to scrapping the Bulleids. However, their boilers were truly excellent and common sense prevailed. All of the Merchant Navies and substantial numbers of their smaller sisters were rebuilt with Walschaerts' gear, losing the casing in the process, and reappeared in much more conventional guise. The rebuilds undoubtedly lost some of the flair of the original design, but proved to be much more reliable.

Bulleid had moved to other things, including the *Leader*, a double-ended tank engine with two self-contained three-axle power bogies driven by what was effectively a Paget engine in miniature. Trials failed to demonstrate an obvious superiority over conventional locomotives, and the project was abandoned before attempts to rectify the worst faults could be made: the 'Leader concept' was far too radical to be pursued at a time when a proposal to abandon steam traction was being considered. Bulleid subsequently moved to the Irish Republic, where he developed a turf-burning locomotive for the CIE public transport company.

By the 1950s, the writing was on the wall for the steam railway locomotive. The desire of modernisers to be rid of what was perceived as old technology impressed their political masters strongly enough for the cull to begin. The French simply disregarded everything Chapelon had achieved, scrapping his impressive prototypes; and British Railways had barely begun to run a series of standard locomotives developed from the comparative 'Interchange Trials' of 1948, ranging from 2-6-0 to 2-10-0, before they were withdrawn in favour of the first generation of diesels. Some of these steam locomotives were scrapped with barely seven years of service behind them: a criminal waste of resources, especially as the earliest diesels performed much worse than their proponents had claimed.

Robert Arthur 'Robin' Riddles was born in 1892. If André Chapelon is regarded as the visionary of steam locomotive design, Riddles may be the ultimate pragmatist. Charged with imposing standardisation on the post-1948 nationalised British Railways, Riddles attempted to introduce a series of simple designs to enginemen who were so set in their ways that no self-respecting GWR driver would have anything good to say about an LMR, LNER or SR rival. A series of lengthy trials showed that many older locomotives had excellent features, but there were far too many types and, in addition, the ravages of war had prevented maintenance on peacetime scales. The BR Standard locomotives included two types each of 2-6-0, 2-6-2T and 4-6-0; a 2-6-4T; two classes of two-cylinder Pacifics (4-6-2); a single three-cylinder 4-6-2 with Caprotti valve gear; and a 2-10-0. No sooner had these locomotives entered service in large numbers than British Railways decided to dispense with steam entirely. They were characterised by simplicity, with outside cylinders and raised running plates for easy access to the bearings, but were doomed to be scrapped with a large part of their useful life still to come. Riddles retired, somewhat disillusioned by the haste with which his locomotives had been discarded, and died in 1983.

Though the British Standard classes were two-cylinder simple expansion designs (excepting the solitary three-cylinder *Duke of Gloucester*), departing greatly from the traditions of keeping every accessory well hidden in case the aesthetics were compromised, they proved to be capable performers once teething troubles had been largely overcome. Among the most impressive was the 2-10-0 9F heavy freight engine, of a type unseen on British metals until the end of the Second World War, which proved to be not only an excellent haulier but also (despite wheels of only 5ft diameter) surprisingly fast. Locomotives of this type could haul loads of up to 900 tons at 35mph on level track, but are said to have been capable of 90mph on heavy excursion trains! High speeds wore the motion and the bearings too rapidly, and so a directive restricting maximum speed had to be circulated.

Some locomotive crews were determined to end in a blaze of glory, and the last authenticated speed of more than 100mph in Britain was recorded on 26 June 1967 by rebuilt Merchant Navy-Class 4-6-2 No. 35003 *Royal Mail*, which, with a light train, ran a mile in thirty-four seconds – a speed of nearly 106mph.

Though steam no longer has a place on the world's main lines, enthusiasts have ensured that the story continues. Ambitious plans to recreate lost classes have been made, and, in some cases, have already been completed. The sole British Railways-Class 8P, *Duke of Gloucester*, has not only been returned to working condition (which involved recreating the Caprotti valve gear) but has also had improvements a wastefully short career had denied. Consequently, the locomotive performed better in preservation than it had ever done in service and is generally reckoned to be the most efficient to have run on British railways.

The Tornado Trust has successfully built a Peppercorn A3 Pacific, which is allowed to reach 90mph on the track. Now an attempt is to be made to recreate the Gresley P2 2-8-2, the most powerful rigid frame locomotive ever to run in the United Kingdom. Consideration has also been given to the production of ten British Standard 2-6-2 3MT tank engines, which are ideally suited to some of the heritage lines. The recreation of the Pennsylvania Railroad 4-4÷4-4 T1, which would be an awesome sight at full speed, has already been mentioned. Where next? Who can tell …

4

STEAM LOCOMOTIVE CLASSIFICATION

Locomotives are classed on the basis of driven axles, no matter how they may be constructed. Consequently, the list intermixes rigid wheelbase and articulated designs indiscriminately, regardless of whether the axles are split into separately driven groups, and so Garratts may appear alongside Mallets, Fairlies, Shays, and a host of other unconventional 'one-off' designs. The basis of the notation is that the wheels on each axle should be in contact with the rails, even if the wheel is in fact a toothed cog acting as – or apparently duplicating – a wheel; thus the original Blenkinsop rack-rail locomotive classifies here as 2-2-2 (it is customarily listed as an 0-4-0) and the jackshafts popularised briefly in the middle of the nineteenth century (and again on many light logging or 'pole-road' designs of 1880–1914) have been ignored. Restrictions on space mean that each entry tends to concentrate on 'first' and 'last', and not necessarily the many locomotives in between. In addition, entries can have an understandable British bias. Minor changes have been made here to improve the Whyte notation. Rigid-wheelbase locomotives are simply '0-6-0'; divided drive and duplex locomotives are '0-6÷6-0'; Mallets and similarly articulated designs are '0-6=6-0'; and Garratts, effectively two separate engines, are '0-6-0+0-6-0'.

One Driving Axle

Oo

0-2-2, or A1

Alternative European classifications: 011 or 1/2.

Owing to the absence of any guidance to the driving wheels, which promoted unsteadiness at speed, the arrangement was short-lived. Rarely encountered outside Britain, it was soon replaced by the 2-2-2, which, by virtue of a longer wheelbase and a leading axle, ran much more steadily.

The first locomotive of this design was Stephenson's *Rocket*, built for the Liverpool & Manchester Railway in 1829. The layout was perpetuated only by the locomotive's immediate L&MR successors. These culminated in *Northumbrian*, with the cylinders, previously diagonal, brought down almost horizontally to prevent the thrust of the pistons trying to lift the machine off the track at speed. *Rocket*'s own cylinders were lowered from 38 degrees to 8 degrees above horizontal following a serious derailment early in 1831.

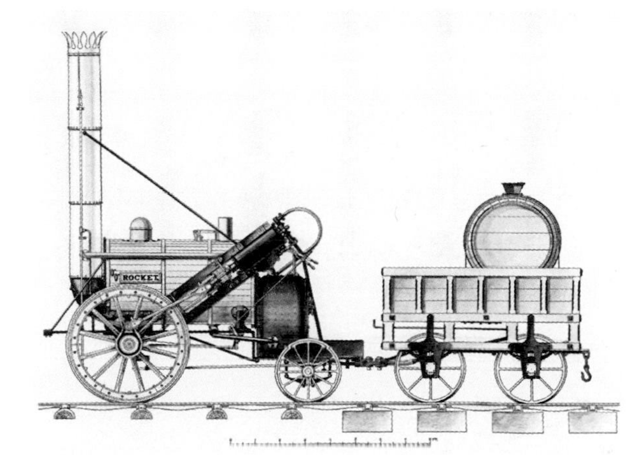

Rocket of 1829. This drawing shows the locomotive 'as built', with the cylinders angled at about 38 degrees to the track. They became near-horizontal almost as soon as *Rocket* had entered traffic. (Courtesy of the Science Museum; Crown Copyright)

Ooo

0-2-4, or A2

Alternative European classifications: 012 or 1/3.

Owing to the absence of any guidance to the driving wheels, which promoted unsteadiness at speed, the arrangement was short-lived. It was also very uncommon, chiefly restricted to locomotives with an unconventional drive train. Typical of the 0-2-4 Class was 9½-ton *Earl of Airlie*, built by J. & C. Carmichael of Dundee for the 4ft 6in-gauge Dundee & Newtyle Railway in 1833. Two 11 × 18in cylinders were placed vertically alongside the boiler, where they drove the 5ft-diameter front wheels through bell cranks. The bogie was the first to be fitted to an engine running in public service in Britain, though a swivelling truck, patented by William Chapman in 1812, seems to have been used on one of the Wylam colliery engines in *c.* 1815.

oO

2-2-0, or A1

Alternative European classifications: 110 or 1/2.

Another of the earliest wheel arrangements, this was confined to compact inside cylinder engines pioneered by Robert Stephenson's *Planet* of 1830. Though the 2-2-0 layout had the merit of guidance for the front wheels, it prevented the firebox being extended backward and was soon superseded by the 2-2-2. Delivered to the Liverpool & Manchester Railway in September 1830, *Planet* was driven by two 11 × 16in cylinders (between the frames beneath the smokebox) and had outside bearings; the frame was the so-called flitched or 'sandwich' type. The first outside cylindered example was *Vauxhall*, built by George Forrester. Delivered in 1834 to the Dublin & Kingstown Railway, the engine proved to be unstable and was reconstructed in 1836 as a 2-2-2. It had 5ft driving wheels and 11 × 18in cylinders.

The Dublin & Kingstown Railway 2-2-0 *Hibernia*, with bell-crank drive. (Engraving originally published in *The Engineer*, 1883)

This model, dating from the 1840s, depicts the outside-cylinder 2-2-0 of the day. Note how the crosshead is supported beneath the frame, and the outside cylinders. Locomotives of this type, with their ultra-short wheelbase and widely spaced cylinders, ran unsteadily and were soon eclipsed by inside-cylinder designs driving on to a crank axle. (Formerly in the collection of the British Engineerium, Hove, England)

oOo

2-2-2, or 1A1

Alternative European classifications: 111 or 1/3.

2-2-2s soon replaced the 0-2-2 and 2-2-0 types for passenger work. The addition of a trailing carrying axle allowed the firebox to be extended behind the driving axle, which in turn allowed the box floor to be lowered and an adequate grate to be fitted. Beginning with Stephenson's *Patentee*, 2-2-2 engines remained popular in Britain into the 1880s even though they were gradually overtaken by bogie patterns (e.g. 4-2-2) for high-speed work.

Developed by Robert Stephenson in the early 1830s, *Patentee* was built for the Liverpool & Manchester Railway in 1833. The engine was driven by two 12 × 18in cylinders inside the flitched frames and had 5ft-diameter driving wheels. The first 2-2-2 with outside cylinders was a long boiler example with 14 × 22in cylinders and 5ft 6in drivers, supplied by Robert Stephenson & Company to the Yarmouth & Norwich Railway in 1844. The last 2-2-2s to be made were apparently Tk Class 5ft gauge tank engines made in 1931–33 by Henschel & Sohn of Kassel for the Latvian state railways. Intended for ultra light passenger duties, they had two outside cylinders measuring 32 × 52cm (12⅝ × 20½), 1.5m (4ft 11in) driving wheels, and a weight of 37.2 tonnes in running order. Boiler pressure was 14at (206 psi).

The livery of the London, Brighton & South Coast Railway – pictured here on the Stroudley 2-2-2 Class 'G' No. 332 *Shanklin*, built in Brighton in July 1881 – must have presented cleaners with a nightmare, especially as the interior of the cab was cream! The locomotive, which had 6ft 6in driving wheels and two inside 17 × 24in cylinders, was withdrawn in March 1910. (From a pre-1914 postcard by F. Moore)

oOo+ooo

2-2-2+6, or 1A13

Alternative European classifications: 1113 or 1/5.

One of the most extreme of the first generation of steam locomotives was *Hurricane*, built for the Great Western Railway by R. & W. Hawthorn of Newcastle upon Tyne to the orders of Isambard Kingdom Brunel. Patented in 1836 by Thomas Harrison, *Hurricane* and near-sister *Thunderer* (with two-axle drive) had two 16 × 20in inside cylinders and running gear on one frame and the boiler on another, relying on ball joint connectors to convey steam from the boiler to the cylinders. *Hurricane* weighed 23 tons without the conventional four-wheel tender. The loading of the driving axle was about 6 tons of the 11 contributed by the drive carriage. The 10ft-diameter wheels are the largest known to have been fitted to a railway locomotive.

oOoo

2-2-4, or 1A2

Alternative European classifications: 112 or 1/4.

Another of the rarer arrangements, this was generally confined to tank engines built in the middle of the nineteenth century with a bunker (and often a well tank) carried behind the cab. Some of the earliest examples had fixed axles, but a trailing bogie was standardised rapidly. The first 2-2-4 seems to have been built in 1859 by Robert Stephenson & Company for the Viceroy of Egypt. The 5ft-diameter driving wheels were driven by two 8 × 14in cylinders inside the frames, and a canopied coach was attached behind the cab. Better known were the numerous 2-2-4T locomotives of the North Eastern Railway, most of which were converted from the Fletcher 0-4-4T BTP Class in the early 1900s. Subsequently Class X3 of the London & North Eastern Railway, they had two inside 17 × 22in cylinders and 6ft 1¼in driving wheels.

2-2-4T *Aerolite* of the North Eastern Railway, outshopped from the Gateshead Works of the North Eastern Railway in 1869, as a 2-2-2WT, was a much-rebuilt machine. Note the outside frames and bearings to the leading and driving axles, which give a clue to antiquity. The locomotive lasted in service until 1933, latterly pulling the chief mechanical engineer's saloon.

ooO

4-2-0, or 2A

Alternative European classifications: 210 or 1/3.

The long boiler 4-2-0 suffered some of the disadvantages of the 2-2-0; even though extending the boiler compensated partly for the additional mass of a firebox placed customarily behind the driven axle. This type of

A 4-2-0 1840s long boiler-type locomotive of the type associated with Robert Stephenson. Note that the driving axle passes transversely through the frame ahead of the firebox. This shortened the wheelbase compared with the Cramptons, which, combined with outside cylinders, tended to make the engines unsteady. Their period in vogue was short. This 1864-vintage model was once in the British Engineerium collection.

4-2-0 often rode unsteadily at speed; 4-2-0 Cramptons, however, with the drive axle behind the firebox, were particularly speedy even though their haulage capacity was somewhat limited.

The first 4-2-0 was *Experiment*, designed by John Jervis (1795–1865) and built in 1832 for the Mohawk & Hudson Railroad by the West Point Foundry, New York. The first to be seen in Britain were American Norris-type engines imported by the Birmingham & Gloucester Railway in 1840–41, and a few comparable engines built locally. These all had outside cylinders. Inside-cylinder long boiler locomotives were introduced in 1842 by Robert Stephenson & Company, the earliest being supplied near simultaneously to the York & North Midland and North & Eastern railways. The North & Eastern *North Star* was typical, with 5ft 6in driving wheels and cylinders measuring 14 × 20in. Very few locomotives of this particular notation were made after 1850.

ooOo

4-2-2, or 2A1

Alternative European classifications: 211 or 1/4.
Popular names: Single-Wheeler, Spinner.

Locomotives embodying this particular wheel arrangement were very popular, particularly in Britain where single wheelers maintained a monopoly on high-speed express passenger work until loads became too great in the 1880s. They were generally replaced either by 4-4-0 or 4-4-2 designs.

The first to embody the 4-2-2 wheel arrangement was *Great Western*, built as a broad-gauge 2-2-2 in the Swindon Works of the Great Western Railway in 1846. After fracturing the leading axle on trials, the engine was rebuilt successfully as a 4-2-2. It had 8ft driving wheels and two inside cylinders measuring 18 × 24in. The carrying axles were all fixed, but the altered *Great Western* proved to be fast and powerful. The first 4-2-2 to carry a bogie was the standard-gauge Single-Wheeler designed by Archibald Sturrock for the Great Northern Railway and built by R. & W. Hawthorn & Company in 1853. The GNR engine had 7ft 6in driving wheels, 17 × 24in cylinders inside the frames, and weighed about 37½ tons in working order.

Though single wheelers were never popular in the USA, the Baldwin Locomotive Company made two Vauclain Compound 'Camelbacks' in 1895–96 to haul lightweight high-speed trains on the Philadelphia & Reading Railroad. This they did successfully until the early 1900s. Among the last single wheelers were those supplied in 1910 to the Pekin–Kalgan and Shanghai–Nanking railways in China, the locomotives being made by Armstrong Whitworth & Co. Ltd and Kerr, Stuart & Co. Ltd respectively. Kerr, Stuart examples had 7ft driving wheels, two external 18 × 26in cylinders, and a boiler pressure of 180psi. Stephenson valve gear was used.

These 4-2-4 tank engines were built by Rothwell & Co. for the broad-gauge Bristol & Exeter Railway to the designs of John Pearson. One is said to have reached 81.8mph – a tremendous speed for its day, perhaps a world record. The 8ft 10in-diameter driving wheels were flangeless, but the locomotives had short lives; they became 4-2-2 tender engines after the B&ER was incorporated into the Great Western Railway.

oo[:o:]Oo

4-2-2-2, or 2AA1

Alternative European classifications: 2111 or 2/5.
Delivered to the Bavarian State Railway by Krauss & Co. of München in November 1895, derived from the standard B XI Class 4-4-0, this unique locomotive (pictured on page 17) was tested extensively. The machine was basically a two-cylinder compound 4-2-2, with a high-pressure cylinder measuring 38.5 × 61cm (15⅛ × 24in) and a low-pressure unit of 61 × 61cm (24 × 24in). However, it could be converted to 4-2-2-2, as an auxiliary drive axle behind the bogie could be engaged at will (by supplying steam to a 'lowering cylinder') to help adhesion. The booster was driven by two 29 × 46cm (11⅜ × 18⅛in) cylinders, one placed directly below each of the main drive cylinders. The locomotive operated at a boiler pressure of 13at (191psi) and weighed 52.6 tonnes in working order; the three-axle tender contributed an additional 34.7 tonnes.

Trials showed that the system allowed the locomotive to pull heavier loads up gradients, but also that the gains were not worth the additional complexity. As train loads were increasing rapidly, so No. 1400 was abandoned – though the idea of a booster was perpetuated in a larger version based on a 4-4-2 (see pages 92–3).

ooOoo

4-2-4, or 2A2

Alternative European classifications: 212 or 1/5.
Locomotives embodying this wheel arrangement were very rare, except for the high-speed tank engines described below. Most were made for track inspection and similar specialised uses, owing to the very low adhesion weight which prevented heavy loads being hauled.

The first engines of this type were a series of broad-gauge tank engines (pictured on the previous page) built by Rothwell & Company of Bolton for the Bristol & Exeter Railway in 1853–54. Designed by John Pearson, they were distinguished by inside cylinders measuring 16½ × 24in and flangeless driving wheels with a diameter of 8ft 10in (sometimes listed as '9ft'). They were exceptionally fast and rode very steadily, but were eventually converted by the GWR to something much more mundane.

6-2-0 *John Stevens* was one of seven near-identical locomotives built in 1848–49 by R. Norris & Sons for the Camden & Amboy Railroad. The 8ft-diameter driving wheel and small-diameter boiler placed as low as possible accorded with the ideas of Crampton, but the lack of adhesion proved to be a weakness.

oooO

6-2-0, or 3A

Alternative European classifications: 310 or 1/4.
This notation was unique to the Cramptons, with their single driving axles mounted behind the firebox – preserving balance at the expense of rigidity and poor adhesion. This prevented heavy loads being hauled.

The best known 6-2-0 is undoubtedly *Liverpool*, built by Bury, Curtis & Kennedy in 1848 for the London & North Western Railway. The engine had 8ft-diameter driving wheels, two 18 × 24in cylinders outside the flitched frames, and weighed about 35 tons. Experiments were abandoned when *Liverpool*, though capable of more than 70mph, proved to be too heavy for contemporary permanent way. The carrying axles were all mounted rigidly in the frame. A few similar locomotives were made in France (by Cail, amongst others) and also by Norris Brothers of Philadelphia, apparently beginning in 1848 with No. 30 *John Stevens* of the Camden & Amboy Railroad. The Norris Crampton preserved the classic form, with a low-slung boiler and unusually large driving wheels, but the leading wheels were all carried in a pivoting bogie.

Two Driving Axles

OO

0-4-0, or B

Alternative European classifications: 020 or 1/2.
The first locomotive to feature driving axles coupled with rods was made *c.* 1816 by George Stephenson, though the rods were carried internally. *Locomotion No. 1*, built in 1825 for the Stockton & Darlington Railway,

Protected by US Patent 277994 granted in 1883 to William Cole of Montgomery, Alabama (and made by Adams & Price of Nashville, Tennessee), this Pole Road 0-4-0T was destined for track made of rough cut logs laid end to end. Note the concave-tread wheels.

This model 0-4-0 probably dates from the 1840s. Made as a toy, it reflects the primitive nature of the earliest generation of locomotive engines. Note 'REVERSING BOX' on the smokebox side. Pulling the lever on the firebox operates a primitive form of gab-type valve gear to reverse the motion.

is generally agreed to have been the first to feature wheels coupled with external rods. The first locomotive to be successfully run in the USA was an 0-4-0 named *Stourbridge Lion*, which had been built by Foster, Rastrick & Company. *Liverpool*, designed by Edward Bury and built in the summer of 1830 by Bury & Kennedy for the Liverpool & Manchester Railway, was the first to feature inside bearings, bar frames, and inside cylinders driving a cranked axle. The engine had 6ft-diameter coupled wheels and two cylinders measuring 12 × 18in. The first railway engines designed specifically for goods traffic and banking work were the 0-4-0s *Samson* and *Goliath*, built by Robert Stephenson & Company for the Liverpool & Manchester Railway in 1831. They had 4ft 6in coupled wheels, two 14 × 16in cylinders – carried inside the frame – and weighed 10 tons in working order (engine only).

Locomotives of this type came to be associated with slow-speed goods traffic, owing to the lack of a guiding truck or bogie, but were soon replaced by the 0-6-0 and 0-8-0 for main-line use. However, thousands of 0-4-0 tank engines were used for shunting and industrial use until the end of steam.

OOo

0-4-2, or B1

Alternative European classifications: 021 or 2/3.

Most of these were restricted to slow-speed mixed-traffic duties, owing to the absence of a leading truck or bogie. They were displaced from passenger working by 2-4-0s or 4-4-0s as train loads grew, and from freight by the 0-6-0.

The earliest 0-4-2 was a goods engine for the Leicester & Swannington Railway, delivered in December 1833. Built by Robert Stephenson & Company, it had 4ft 6in coupled wheels and two 14 × 18in cylinders inside flitched frames. The engine weighed about 14 tons in working order. William Stroudley of the London, Brighton & South Coast Railway subsequently built the D1 or Gladstone Class of 0-4-2 express engines in the 1880s. Sceptics suggested that the 6ft 6in coupled wheels would leave the tracks at the earliest opportunity, though the designer countered that the large diameter of the leading wheels gave a greater flange contact with the rails. He claimed that this was particularly advantageous on points and crossovers, and the Gladstones soon proved to be swift and steady runners (if expensive to maintain). They were popular with enginemen largely because the cab, supported by a trailing axle, was isolated from the pounding of the cranks. They lasted into the 1920s.

The basic layout was also perpetuated on the 14xx Class of 0-4-2T tank engines built in Swindon Works for the Great Western Railway, which lasted virtually until the end of steam traction on British Railways. Particularly popular on 'push pull' trains in latter days, these tank locomotives were reputedly capable of 70mph. The layout was also popular on small tender engines, particularly narrow gauge, as the trailing axle allowed the cab to be supported and improved riding qualities.

OOoo

0-4-4, or B2

Alternative European classifications: 022 or 2/4.

This layout was widely applied in Britain to tank engines, particularly if destined for suburban passenger duties. It was also popular on their North American equivalent – the so-called Forney Tanks developed for the elevated railways that ran in many large cities. However, few 0-4-4 tank engines were made after 1930 owing to the increased popularity of the 2-6-2 and 2-6-4 patterns that not only developed greater power but also ran better at speed.

Left: 0-4-4T No. 111 of the Lancashire & Yorkshire Railway was built in 1877 by Kitson & Co. of Leeds to the design of W. Barton Wright, the railway's locomotive superintendent. It was used initially on passenger trains from Manchester to the Leeds area.

The first engines of this type were designed by James Cudworth and made by Brassey & Company of Birkenhead for the South Eastern Railway. Delivered in 1866, they were basically long bunker derivatives of an earlier 0-4-2T. They had 15 × 20in inside cylinders, 5ft 7in driving wheels and weighed 33 tons 14cwt in working order. Virtually every other British railway followed this lead, almost always with inside cylinders.

Above: One of the celebrated Forney Tanks, this 0-4-4T was built in 1876 by the Schenectady Locomotive Works for the New York & Harlem Railroad. Engines of this type were generally used on 'mass movement' passenger railways with short distances between stations.

OO+oo

0-4-4, or B2

Alternative European classifications: 022, 2/4.
Most of these, which bear a considerable external resemblance to the standard 0-4-4T (q.v.), are Single Fairlies. The driving and carrying axles, each in a separate group, are pivoted independently. The first examples were designed for the 5ft 3in-gauge Great Southern & Western Railway of Ireland by Alexander McConnell. Built in the railway's Inchicore (Dublin) works in 1869–70, they had 15 × 20in cylinders and 5ft 7½in driving wheels. Essentially similar Mason Fairlie locomotives were built in the USA.

OoooO

0-2-6-2-0, or A3A

Alternative European classifications: 131, 2/5.
This extraordinary design (pictured on page 34) was limited to a class of express passenger engines made for the Chemins de Fer du Nord (France) to the designs of Jules Petiet. Eight were made by Gouin of Paris in 1862–63, one being exhibited at the London Exhibition of 1862. They had a pair of cylinders at each end of the frame, separated by three small-wheel carrying axles, and by a chimney carried horizontally backward to exhaust above the cab. A six-axle goods engine of this type was also made.

OO+ooo

0-4-6, or B3

Alternative European classifications: 032, 2/5.

The first of this notation was *Thunderer*, built to the patent of Joseph Harrison, which was effectively an 0-4-0 with 6ft-diameter coupled wheels mated with a boiler carried on a separate three-axle carriage.

Gearing was used to increase the speed of rotation. The other locomotives of this notation were Mason Fairlies, with a single driving bogie. The three-axle carrying bogie was needed to support an extended fuel bunker.

oO÷O

2-2÷2-0, or 1AA

Alternative European classifications: 1110 or 2/3.

This wheel arrangement has been confined to a handful of split-drive designs, briefly popular in Britain and France from the 1880s until 1900. Though they all give the appearance of 2-4-0 notation, the driving wheels were not coupled together in an attempt to reduce friction. The first of Francis W. Webb's three-cylinder compounds was *Experiment*, completed in the Crewe works of the London & North Western Railway in April 1882. The engine had a single low-pressure cylinder inside the frames, with a 26in bore and a 24in stroke, and two outside high-pressure cylinders of 13 × 24in. Driving wheels were 6ft 9in, boiler pressure was 150psi and the engine weighed 37 tons 13cwt in running order. It was followed by the Dreadnought and Teutonic classes, introduced in 1884 and 1889 respectively. One Webb-type compound – complete with pilot (cowcatcher) and large cab – was supplied in the 1890s to the Pennsylvania Railroad, where it ran for a few years without distinguishing itself.

The first compound developed by Alfred de Glehn took 2-2÷2-0 form, but was successful only when changed by Gaston du Bousquet to conventional 4-4-0 form. Once this had been done, the locomotives created a lasting reputation for themselves and, largely in retrospect, also for de Glehn.

oOO

2-4-0, or 1B

Alternative European classifications: 110 or 2/3.

This was an extended version of the 2-2-0, relying on a long narrow firebox carried between the rearmost driving wheels to maintain balance (cf., 4-2-0). The leading axle helped to guide the coupled wheels and improved riding at speed. The 2-4-0 was popular in Britain, particularly as a passenger or mixed-traffic haulier, but eventually lost favour to the 4-4-0. It was also comparatively common in Europe, where many original locomotives, converted to 2-4-2 by the addition of a trailing carrying axle, survived into the 1890s. However, it was rarely seen in the USA other than on switchers, logging and industrial engines.

The earliest known 2-4-0s were supplied by Robert Stephenson & Company in 1837, two to the USA and two to the Paris & Versailles railway. They had 4ft 6in coupled wheels and two inside cylinders measuring 15 × 18in. The first locomotives of this particular notation to run in Britain

were purchased by the Great North of England Railway in 1839–40. The earliest known outside cylinder 2-4-0 was a long boiler pattern supplied by Robert Stephenson & Company to the York, Newcastle & Berwick Railway in 1845.

2-4-0 *Wetzlar* of the Württemberg state railway, delivered from the Esslingen factory of Emil Kessler in 1875, typified the long boiler designs that were still favoured in central Europe. Short wheelbase and outside cylinders did not suit them to high speed.

oO÷Oo

2-2÷2-2, or 1AA1

Alternative European classifications: 1111 or 2/4.

This wheel arrangement has been confined to a handful of split-drive designs credited to Francis Webb of the London & North Western Railway, briefly popular in Britain at the end of the nineteenth century. Though they all give the appearance of 2-4-2 notation, the driving wheels were not coupled together in an attempt to reduce friction. None of them were entirely successful, though this was due at least as much to poor boiler design as to the divided drive.

The first locomotive of this type was *Greater Britain*, completed in the LNWR works in Crewe in October 1891. This machine had 7ft 1in driving wheels and weighed 52 tons 2cwt without the tender. The large low-pressure inside cylinder had a bore of 30in and a stroke of 24in; the two external high-pressure cylinders measured 15 × 24in; and boiler pressure was 175psi. The last of the twenty 2-2÷2-2s to be made was *William Siemens* of the John Hick Class, completed in Crewe in 1898. This locomotive was similar to *Greater Britain* but had 6ft 3in driving wheels.

Webb compound 2-2÷2-2 *Queen Empress* of the London & North Western Railway (Crewe works, 1891) gained a gold medal from the Chicago World Fair of 1893. The driving wheels had a diameter of 7ft 1in, two 15 × 24in high-pressure cylinders were mounted externally and there was a single 30 × 24in low-pressure cylinder between the frames.

oOOo

2-4-2, or 1B1

Alternative European classifications: 121 or 2/4.
Popular name: Columbia.
Commonly used in the nineteenth century on tank engines that were required to run at speed in either direction, this arrangement is not as commonly encountered on tender engines.

The earliest 2-4-2s were tender engines built by Robert Stephenson & Company for the Belgian-owned Great Luxembourg Railway. Delivered in 1860, the prototype had 15 × 22in cylinders and 5ft 6in driving wheels. *White Raven*, a tank engine completed in 1863 by Cross & Company of St Helens, was the first to incorporate radial axle boxes. Built for the St Helens Railway, *White Raven* had 5ft 1in coupled wheels, 15 × 20in inside cylinders, and weighed 40 tons 16cwt in working order.

Columbia, a Vauclain Compound with 7ft 0¼in driving wheels, was exhibited by the Baldwin Locomotive Company at the Columbian Exposition (Chicago) in 1893. The engine weighed 91 tons 14cwt with its tender, was 63ft 4in long and gave this particular category its popular name. Baldwin also exhibited a Camelback 2-4-2 built for fast passenger service on the Philadelphia & Reading Railroad.

The last British 2-4-2T locomotives were apparently made for the Great Eastern Railway, though others were made in Europe until the 1940s. These included *Postdampflokomotiven* (with luggage compartments) built for the Austrian railways in 1934–38 by Wiener Lokomotiv-Fabriks AG of Floridsdorf. They had two outside 29 × 57cm (11⅜ × 22½in) cylinders, boilers pressed to 16at (235psi) and 1.45m- (4ft 9in-) diameter driving wheels. Overall length was 11.2m (36ft 9in), weight in running order being approximately 41.8 tonnes. *Postdampflokomotiven* were intended to run lightly loaded services – sometimes with a coach or wagon, sometimes ‘light engine’ – on rural lines, so axle loading was kept as low as possible.

This 2-4-2T of the London & North Western Railway was nominally a rebuild of a 2-4-0 Precursor, which had been converted in Crewe works from 1890 onward. The comparatively large driving wheels suited Precursor tanks largely to local passenger trains.

oOOoo

2-4-4, or 1B2

Alternative European classifications: 122 or 2/5.
This uncommon arrangement will be found on tank engines with the bunker extending backward far enough to require greater support than a single carrying axle could provide. One locomotive of this type was created for the Belfast & Northern Counties Railway in 1931 by rebuilding a two-cylinder compound 2-4-2T No. 110 supplied in 1892 by Beyer, Peacock & Co. Ltd. The trailing bogie supported the new boiler, which had a Belpaire firebox.

Made by the Rogers Locomotive Works, 2-4-4T No. 213 ran on the Illinois Central Railroad.

oOO+oo

2-4-4, or 1B2

Alternative European classifications: 122, 2/5.
These were all Mason Fairlies, fitted with a leading axle to improve high-speed performance.

oOO+ooo

2-4-6, or 1B3

Alternative European classifications: 123, 2/6.
The rarely encountered Mason Fairlie locomotives of this type, often intended for passenger or mixed-traffic duties, generally had an extended bunker.

Unique to the USA, Mason Fairlie Double Truck locomotives were briefly popular before loads became too great. This 2-4-6T was made in 1885 for the Providence, Warren & Bristol Railroad.

oOo+oOo

2-2-2+2-2-2, or 1A1+1A1

Alternative European classifications: 111+111, 2/6.
This extraordinary arrangement was confined to six Beyer, Peacock & Company 3ft 6in-gauge Garratts made in 1924–25 for South Africa. The members of the GE Class were apparently intended to provide a combination of adequate hauling capacity and fast running on poorly laid track, but the experiment was never repeated.

ooO÷O

4-2÷2-0, or 2AA

Alternative European classifications: 2110 or 2/4.
This wheel arrangement was confined to a split-drive three-cylinder compound locomotive designed by Francis Webb for the London & North Western Railway, and four-cylinder machines promoted by Dugald Drummond of the London & South Western. Though the locomotives had the appearance of 4-4-0s, their driving wheels were not coupled together in an attempt to reduce friction.

The first of the type was LNWR tank engine No. 2063, converted from a 4-4-0 Metropolitan locomotive originally made by Beyer, Peacock & Company of Manchester. The compound re-entered service in June 1884, with one inside cylinder measuring 26 × 24in (low pressure) and two outside cylinders of 13 × 24in (high pressure). Boiler pressure was 150psi, the diameter of the driving wheels was 5ft 9in and the locomotive weighed 46 tons 17cwt in working order.

Drummond's engines were all four-cylinder simple expansion designs, with two inside cylinders driving the front axle and two outside cylinders driving the rear axle. The first, T7 Class No. 720, was completed in Nine Elms works in 1897. The driving wheels were 6ft 7in, the cylinders measured 16½ × 26in (soon reduced to 15 × 26in) and boiler pressure was 175psi. The locomotive and tender were 60ft 1¼in overall and weighed 103 tons 11 cwt in running order. No. 720 was followed in 1901 by five similar locomotives of the E10 Class, but none was successful.

ooOO

4-4-0, or 2B

Alternative European classifications: 220 or 2/4.
Popular name: Eight-Wheeler or American.
Beginning in 1833, when a locomotive of this notation was sent to the USA from the Stephenson factory in Newcastle upon Tyne, these engines were very popular in the nineteenth century; the bogie enabled curves to be taken at speed, suiting them to passenger services. Superseded elsewhere by larger machines by 1900, the 4-4-0 remained widespread in Britain into the 1930s.

Patented in the USA in February 1836 by Henry Campbell of the Philadelphia, Germantown & Norristown Railroad, the first indigenous 4-4-0 to incorporate a bogie was completed by Matthias Baldwin of Philadelphia in May 1837; the perfected design, with three-point suspension, appeared in 1839.

The first 4-4-0 to run in Britain was adapted by the Birmingham & Gloucester Railway *c.* 1846, from a Philadelphia-built Norris 4-2-0. The first to be built in Britain were bogie saddle tanks constructed in the Great Western Railway's Swindon shops in 1847. They had 6ft-diameter coupled

The Eight-Wheeler or 4-4-0 became known as the American, owing to its popularity in the USA. The *General* was particularly famous. Built in December 1855 by Rogers, Ketchum & Grosvenor of Paterson, New Jersey, for the 5ft-gauge Western & Atlantic Railroad, the locomotive was hijacked in April 1862 by James Andrews and his collaborators in an attempt to wreck the Confederate railroad network. The story of the chase has been told in several films, but ended in tragedy with the execution of several participants.

4-4-0 *Brougham* was delivered by Robert Stephenson & Co. to the Stockton & Darlington Railway in 1860 and withdrawn in 1888. Note the unusually commodious cab for a British locomotive of this era, which would have been appreciated in a harsh winter.

4-4-0 No. 78 of the Glasgow & South Western Railway, designed by James Manson and built in the company's Kilmarnock shops in 1893, typified the supreme elegance of the nineteenth-century British locomotive. Note the unusual tender, with two fixed axles and a bogie.

This American-type 4-4-0, built in 1895 by the Richmond Locomotive & Machine Works for the Seaboard Air Line, is renowned for a record-breaking run made between Weldon, North Carolina, and Portsmouth, Virginia, on 21 November 1896. Rebuilt as Class G in 1916, No. 540 (then 170) was withdrawn in 1933.

wheels and two 17 × 24in inside cylinders. Bogie tender engines were made in Britain in 1854–58 for the Arica & Tacua and Copiapo & Caldera railways in Chile (by R. & W. Hawthorn and Kitson & Co. respectively), but the first to run on British tracks was *Brougham* of the Stockton & Darlington Railway – one of two locomotives made in 1860 by Robert Stephenson & Company to the designs of William Bouch (1813–76). They had 6ft coupled wheels and 16 × 24in outside cylinders. The first English type 4-4-0 with inside cylinders and inside frames, designed by Thomas Wheatley, was delivered to the North British Railway in 1871. one locomotive of this type was lost when the Tay Bridge collapsed in a gale in 1879.

ooOOo

4-4-2, or 2B1

Alternative European classifications: 221 or 2/5.
Popular name: Atlantic.
The name was coined in 1894 by Jerome Kenley, general manager of the Atlantic Coast Railroad, for locomotives supplied by Baldwin. Created in the late 1880s, these engines were very popular prior to 1914; the bogie enabled curves to be taken at speed, suiting them to passenger services, and the trailing axle allowed a capacious firebox to be fitted. The layout was also popular on tank engines, where the trailing axle was used to support the bunker. Though the Atlantic lost ground after 1900 to the 4-6-2 and 4-6-4 for passenger work, owing to increasing loads, it enjoyed a brief renaissance in the 1930s in the USA, Belgium and elsewhere. The Chicago, 'Milwaukee', St Paul & Pacific Railroad – nowhere near the Atlantic! – designated A Class 4-4-2 machines delivered by Alco in 1937 as 'Milwaukee'.

The earliest 4-4-2T, dating from 1882, was derived by William Adams from a standard London & South Western Railway 4-4-0. The work of Beyer, Peacock & Company of Manchester, it had two outside cylinders (17½ × 24in), 5ft 7in-diameter coupled wheels and a boiler pressure of 150psi. Overall length was 38ft 8¼in, weight in running order being 54 tons 2cwt.

The first tender engine, designed by George Strong for the Lehigh Valley Railroad, was built in 1888 by the Vulcan Iron Works of Wilkes Barr, Pennsylvania. The first Atlantic to be built in Britain was No. 990 *Henry Oakley*, designed for the Great Northern Railway by Henry Ivatt (1851–1923), and delivered from the company's Doncaster works in 1898. The first wide firebox design was Ivatt's No. 251 of 1900.

A renewal of interest in the Atlantic occurred in the mid-1930s, when a special large-wheeled class was introduced with the 'Hiawatha' high-speed service. Perhaps the last to be made were a handful of high-speed Class 12 Atlantics made in 1938–39 by Société John Cockerill of Seraing for the Belgian state railways. These had two inside cylinders (48 × 72cm, 18⅞ × 28⅜in), 2.1m (6ft 10½in) coupled wheels and a boiler pressure of 16at (235psi).

One of the Highflyers of the Lancashire & Yorkshire Railway, 4-4-2 No. 700, was built in Horwich works in 1899 to the designs of John Aspinall. The driving wheels had a diameter of 7ft 3in, and there were two 19 × 26in cylinders between the frames.

LMS 4-4-2T No. 2111 was one of a batch of ten built in Derby works in 1924 to a design that had originated on the London, Tilbury & Southend Railway. The engines were essentially similar to the group, designed by Thomas Whitelegg, which had been ordered from Dübs and Sharp, Stewart of Glasgow in the late 1890s. The 6ft 6in-diameter wheels suited the locomotives to suburban commuter trains running to tight schedules.

o[:o:]oOOo

2-2-2-4-2, or 2AB1

Alternative European classifications: 2121 or 3/6.

Derived from the earlier 4-2-2/4-4-2 dating from 1895, also the work of Krauss & Company of Munich, this interesting departure from standard practice was exhibited at the Paris Exhibition of 1900 and tested extensively on the Bavarian railway system – notably in the Pfalz mountains – in 1901–02. The basis was a conventional Atlantic (4-4-2), but the notation could be converted at will to 2-2-2-4-2 by lowering a steam-controlled supplementary axle between the bogie wheels on to the rails. The goal was short-term boosts to adhesion to overcome steep gradients and adverse weather conditions, but was attained at such complexity that the experiment failed. The engine was a compound, with a single 44 × 66cm (17¼ × 26in) high-pressure cylinder and a low-pressure cylinder measuring 64 × 66cm (25⅞ × 26in); boiler pressure was 14at (206 psi). A 26 × 40cm (10¼ × 15¾in) auxiliary-drive cylinder, lowered on to the rails by steam pressure in two small cylinders attached horizontally to the mainframe, lay beneath each of the main drive units. The engine and tender were 19.07m (62ft 6in) overall and weighed 113.7 tonnes in working order. The machine was the last of its type, as the tender-mounted booster promised more.

The aberrant Krauss 4-4-2 with the steam-powered 'dolly' booster between the bogie wheels (which could make a 6-4-2!).

ooOoo

4-4-4, or 2B2

Alternative European classifications: 222 or 2/6.
Popular name: Reading or Jubilee.
This uncommon arrangement offered poor adhesion in relation to total weight, being confined largely to tank engines with bunkers extending backward far enough to require greater support than a single carrying axle

Seen in the summer of 1932 in London & North Eastern Railway livery, as No. 2154, this Gateshead-built D-Class 4-4-4T had been designed by Vincent Raven for the North Eastern Railway; forty-five of them were built in 1913–22. Large-diameter driving wheels (5ft 9in) and Westinghouse brakes show that the engine was intended to work passenger trains at high speed. Survivors of the class were eventually rebuilt as 4-6-2T (A8 Class).

could provide. However, some highly individualistic tender engines have also been made.

The earliest British 4-4-4T was built in 1896 for the Wirral Railway by Beyer, Peacock & Company of Manchester. It had two outside cylinders measuring 17 × 24in; the driving wheels had a diameter of 5ft 2in, and the boilers were pressed to 160psi. Total weight was 59 tons 16cwt. The most popular British 4-4-4T was introduced on the North Eastern Railway in 1913. Designed by Vincent Raven, forty-five of Class D were made prior to 1922. They had three 16½ × 26in cylinders, 5ft 9in coupled wheels, and boiler pressures of 160psi. Weight in running order was 84 tons 15cwt.

Large-wheel 4-4-4 tender engines were built for the Baltimore & Ohio Railroad in 1934 (*Lady Baltimore*) and for the Canadian Pacific Railway, beginning with Class F 2a in 1936. Designed specifically to haul comparatively lightweight trains at very high speed, the Canadian-built engines had two 17¼ × 28in outside cylinders, 6ft 8in-diameter coupled wheels, and boiler pressures of 300psi. The locomotive/tender unit weighed 209 tons in running order, overall length being 81ft 3in.

Though the 4-4-4 enjoyed a short period in vogue, increases in train loads soon began to tax their capabilities. The obvious replacement was a large wheeled 4-6-4, which offered similar fleetness of foot but better adhesion.

ooOOooo

4-4-6, or 2B3

Alternative European classifications: 223 or 2/7.

This wheel arrangement, which offers a low adhesive weight in relation to total weight, has been largely confined to a handful of American-made large bunker tank engines. The only tender locomotive known to have embodied this strange layout was an experimental French Thuile System high-speed example built by Schneider & Companie of Le Creusot, shown at the Paris Exhibition of 1900. The driver's cab was ahead of the smokebox (for additional details see pages 46–8).

Three Driving Axles

OOO

0-6-0, or C

Alternative European classifications: 030 or 3/3.

Engines of this type were used widely for freight traffic, local passenger services and shunting. Challenged by the 0-8-0, 2-8-0 and 2-10-0 in Britain, they nevertheless remained in service in huge numbers until the end of steam. Comparable designs were relegated speedily to local service in much of Europe (where loads and distances were customarily greater), while they were practically unknown in the USA except as shunting engines or 'yard goats'. The absence of a leading truck or bogie restricted speed.

The earliest 0-6-0 was the 'Steam Elephant', said to have been made *c.* 1816, probably by Chapman & Buddle; and a comparable locomotive by Losh & Stephenson was sent to the Kilmarnock & Troon plateway in 1817. Driven by cylinders placed vertically in the boiler, by way of beams and rods, the axles of these engines were connected with chains.

Royal George, which entered service on the Stockton & Darlington Railway in September 1827, was the first to have six wheels coupled with rigid rods. Designed by Timothy Hackworth and built in his Shildon workshops, the engine had two 11 × 20in vertical cylinders driving the rear axle. The

North Eastern Railway 0-6-0 No. 831, built in 1873 by R. & W. Hawthorn of Newcastle upon Tyne and pictured in 1881, typified the attention to finish and detail paid to even the nineteenth-century British goods locomotive. These particular Hawthorn-made engines were known as 'The Belgians', owing to the use of imported labour to break a strike.

An advertisement placed by Hudswell, Clarke & Co. Ltd, showing one of the company's 0-6-0 tank engines. These were widely used by industrial concerns on standard-gauge and other tracks. (*Engineering*, 1902)

A Belgian outside-frame 0-6-0. A generous loading gauge allowed Belgian designers to prepare locomotives that would have been far too wide to run in Britain!

driving wheels are said to have had a diameter of 4ft. The first to be driven by two inside cylinders (16 × 20in), placed under the smokebox within the flitched frames, was an unnamed outside frame goods engine delivered by Robert Stephenson & Company to the Leicester & Swannington Railway in February 1834. The coupled wheels had a diameter of merely 4ft 6in.

The first 0-6-0 with inside frames and inside bearings was a long boiler pattern delivered by Robert Stephenson & Company to the York & North Midland Railway in 1843. The first outside cylinder/inside frame locomotive was a long boiler example made at Shildon in 1847 for the Stockton & Darlington Railway; designed by William Bouch, it had 16 × 24in cylinders and 4ft-diameter coupled wheels; weight in working order (engine only) was 25 tons 11cwt.

Though the European 0-6-0 was usually comparatively small, some of the 'yard goats' and shunting engines built in the USA were surprisingly large and powerful.

OOOo

0-6-2, or C1

Alternative European classifications, 031 or 3/4.
Wheel arrangements of this type were customarily restricted to tank engines, which were themselves usually versions of 0-6-0 tender engines with the frame, supporting the bunker, extending back behind the cab far enough to require an additional carrying axle. However, a few large firebox tender engines of this type were made in the USA for service as 'yard goats' or on short-haul freight trains.

This Lancashire & Yorkshire 0-6-2T was photographed at Moor Row in July 1932, when it had become LMS No. 11638. Designed by Barton Wright, fourteen of these locomotives were made by Kitson in 1881 and another forty by Dübs in 1882. They had 5ft 1½in driving wheels and 17½ × 26in inside cylinders.

OOO+o

0-6-2, or C1

Alternative European classifications: 031, 3/4.
The only known Single Fairlie of this type – a product of the Vulcan Foundry of Newton le Willows – was delivered to the North Wales Railway in 1915.

OOOoo

0-6-4, or C2

Alternative European classifications: 032 or 3/5.
Rarely found in a tender engine, this layout was more commonly associated with tank engines with a bogie under an extended bunker. The absence of a leading truck or bogie promoted unsteady riding and so 0-6-4 was rapidly superseded by the more efficient 2-6-4 layout.

The earliest 0-6-4 locomotive seems to have been a 1852 vintage modification of Haswell's Semmering trials 0-8-0 *Vindobona*, which had been built by the Wien & Gloggnitz railway workshops in Wien (Vienna) in 1851. This experimental design, however, was soon withdrawn for use as a stationary boiler. Subsequent examples were invariably large bunker tank engines, but large numbers of narrow-gauge locomotives were used in the former German colonies of south-western Africa in the twentieth century.

The first British representative was designed by Alexander McDonnell for the 5ft 3in-gauge Great Southern & Western Railway of Ireland

One of the 0-6-4T 2000-Class Flat Irons, designed by Richard Deeley of the Midland Railway. Forty were made in Derby Works in 1907. They had 5ft 7in driving wheels, two 18½ × 26in cylinders, boiler pressure of 175psi and a weight of 72 tons 8cwt in working order. Unsteady and prone to rough riding at speed, they had all been scrapped by 1938.

in 1876. It had two inside cylinders measuring 18 × 24in, 4ft 6½in driving wheels, and a boiler pressure of 140psi. The engine weighed 46 tons 16cwt in running order.

Nine 0-6-4Ts were delivered in 1886 to the Mersey Railway by Beyer, Peacock & Co. Ltd, but the most numerous were the 'Flat Irons' of the Midland Railway, forty being built in Derby works in 1907 to the designs of Richard Deeley. Some of the largest Forney Tanks, built for elevated railways popular in North American railways, were also of this pattern instead of the more commonly encountered 0-4-4 notation, but the most numerous 0-6-4s were 175 tank engines made by Maschinenfabrik Esslingen for the 3ft 6in-gauge railways of the short-lived South African Republic (ZAR) in 1893–98.

OOO+oo

0-6-4, or C2

Alternative European classifications: 032, 3/5.
The earliest of these Single Fairlie designs was probably *Moel Tryfan*, built by the Vulcan Foundry in 1875 for the North Wales Narrow Gauge Railway. Others were made in 1878–80 for the New Zealand government railways by the Avonside Engine Company of Bristol. The last was a 1ft 11½in-gauge example made in 1909 by the Hunslet Engine Company of Leeds for the North Wales Railway. Single Fairlies proved to have few advantages over rigid-frame designs, apart from the ability to negotiate sharp curves.

OOO+ooo

0-6-6, or C3

Alternative European classifications: 033, 3/6.
Mason Fairlie engines of this type, generally small-wheel examples for goods traffic, were distinguished by fuel bunkers extending backward above the three-axle bogie.

oO÷OO

2-2÷4-0, or 1AB

Alternative European classifications: 1120 or 3/4.

This wheel arrangement has been confined to a split-drive three-cylinder design promoted by Francis Webb of the London & North Western Railway. Though it had the appearance of 2-6-0 notation, only the second and third driving axles were coupled together to reduce friction. The solitary engine of this type was an LNWR 'Compound Goods' tank engine, No. 777, intended for heavy freight service. Built in the company's works in Crewe, the locomotive entered service in December 1887, with one cylinder measuring 30 × 24in (low pressure) and two of 14 × 24in (high pressure). Boiler pressure was 160psi, the diameter of the driving wheels was 5ft and the locomotive weighed 55 tons in running order.

Service soon showed that, though an improvement on what the LNWR was already running, the quirky Webb compound was easily eclipsed by much simpler engines running on other British railways. The experiment was not repeated.

Webb's solitary 2-2÷4-0 compound tank engine, built by the London & North Western Railway in 1887.

oOOO

2-6-0, or 1C

Alternative European classifications: 130 or 3/4.

Popular name: Mogul.

The common name is believed to have come from *Mogul* – completed in 1866 by the Taunton Locomotive Works for the Central Railroad of New Jersey. Baldwin was using the term in sales literature by the early 1870s. Originally conceived for medium-speed freight service, the 2-6-0 eventually developed into a lightweight mixed-traffic haulier.

This 2-6-0 Mogul, *Mt. Washington*, was built in 1879 by the Manchester Locomotive Works for the Boston, Concord & Montreal Railroad.

LMS 2-6-0 'Crab' No. 13020 was made to the designs of George Hughes, chief mechanical engineer of the Lancashire & Yorkshire Railway. Substantial quantities were built in Horwich (70) and Crewe (175) in 1926–32. The nickname is said to have arisen from the movement of the external Walschaerts' valve gear, which was still then comparatively unusual in Britain, but the thrust of the steeply inclined pistons gave the locomotives an unusual motion on the track.

Excepting tank engines, Moguls were generally preferred to the 2-6-2 (Prairie) owing to a better ratio of adhesive to total weight. The first of this type was apparently built in St Petersburg in 1844 by Harrison, Winans & Eastwick, for the St Petersburg–Moscow railway. Rigid frame 2-6-0 locomotives were made by Baldwin and Norris in 1852–53 for service in North America. Another plausible claimant for the title 'the earliest' is *Pawnee*, built by James Millholland for the Philadelphia & Reading Railroad at much the same time. The first to incorporate a pivoting truck, patented by Isaac Bissell in *c.* 1853, was built by Baldwin in 1860 for the Louisville & Nashville Railroad.

The earliest British 2-6-0s were built by Kitson & Company of Leeds in 1866, for service on the 3ft 6in-gauge Queensland Railway in Australia. However, the first to run in Britain was No. 527 *Mogul*, designed by William Adams (1823–1904) and built in 1878 for the Great Eastern Railway by Neilson & Company of Glasgow. No. 527 had two 19 × 26in outside cylinders, 4ft 10in-diameter coupled wheels, and weighed 46 tons 12cwt. Boiler pressure was 140psi.

oOOOo

2-6-2, or 1C1

Alternative European classifications: 131 or 3/5.
Popular name: Prairie.

The addition of a trailing axle often enabled a wide firebox to be fitted to a Mogul (2-6-0), though many Prairies were narrow firebox tank engines with the trailing axle supporting an enlarged bunker. Some wide firebox 2-6-2s proved to be better steamers (and thus better hauliers) than many contemporaneous 4-6-0s, but the absence of a leading bogie was widely considered to restrict speed potential. The 2-6-2 achieved great popularity in Austria–Hungary and Italy, the latter often fitted with Zara trucks, but enjoyed a belated renaissance in Britain between the wars on the London & North Eastern Railway and also in post-war Germany.

The first 2-6-2 was built by the Baldwin Locomotive Works in 1885, for export to New Zealand. Later examples, built in 1900–01 by the Brooks Locomotive Works for the Chicago, Burlington & Quincy Railroad, became popular in the open spaces (prairies) of the American Midwest, creating the generic name.

2-6-2T No. 236 of the Central Pacific Railroad dated from 1882. Locomotives of this type were able to run in either direction with equal facility.

The earliest British representative was a tank engine built in 1887 by Beyer, Peacock & Company for the Mersey Railway. This had two 19½ × 26in outside cylinders, 4ft 7½in-diameter coupled wheels, and a boiler pressed to 150psi. Locomotives of this type weighed about 62 tons 10cwt. The V2 Class 2-6-2 of the London & North Eastern Railway, designed by Nigel Gresley, had three 18½ × 26in cylinders, patented conjugated valve gear, 6ft 2in-diameter coupled wheels, and a boiler pressure of 220psi. The engines alone weighed 93 tons 2cwt in working order; 184 of them were made in 1936–41.

2-6-2T No. 6134, seen here in British Railways livery, was typical of the great numbers of Large Prairie Tanks built in Swindon for the Great Western Railway from the 5100 Class of 1903 onward. The 6100 Class, introduced in 1931, was designed largely for suburban commuter traffic and had 5ft 8in driving wheels.

oOOOoo

2-6-4, or 1C2

Alternative European classifications: 132 or 3/6.
Popular name: Adriatic or Australian.

The basic 2-6-4 layout was most common on tank engines, particularly in Britain where it replaced the unsuccessful 0-6-4 design. The addition of a leading truck improved riding qualities appreciably and increased maximum attainable speed. Consequently, 2-6-4s, especially tank engines became popular for suburban passenger and mixed-traffic duties once concerns about high-speed stability had been allayed. They were not particularly common in Europe, though a few were made by Baldwin and Tampella for service in Finland and the Class 66 was tested by the Deutsches Bundesbahn. A few were made in the USA in the 1920s for suburban passenger traffic.

The first tender engine was introduced in 1901 on the Cape Government Railway in South Africa, though the Class 310 designed for the Austro-Hungarian railways by Karl Gölsdorf, to pull express passenger trains on lightly laid track, is much better known. Tank engines were tried by the Pretoria–Peitersburg Railway as early as 1898, but the first British representative was Class 1B of 1914, designed for the Great Central Railway by John Robinson. The British Railways' 2-6-4T 4MT, however, was by far the most successful.

Tank engines of 2-6-4 notation were very popular in Britain as they allowed the bunker to be extended. This increased range. The locomotive shown, LMS No. 2443 (one of the first batch, built at Derby in 1937), was of a type developed by William Stanier from the LMS Fowler 2-6-4T of 1927 and the Large Prairies of the GWR (Stanier had trained in Swindon); it was destined to be the prototype of the British Railways Standard 4MT 2-6-4T. Two-cylinder No. 2443, by then 42443, was scrapped in 1962.

oOOOooo

2-6-6, or 1C3

Alternative European classifications: 133 or 3/7.

This rare arrangement was featured on Mason Fairlies of the 1880s and on some high-speed tank engines, made in the USA to handle suburban traffic in 1890–1920. The three-axle trailing bogie supported the elongated bunker and well-tank unit. A few 2-6-6T were built by the Brooks Locomotive Works of Dunkirk, New York State, for the Chicago & Northern Pacific Railroad. One was exhibited at the Columbian Exposition, Chicago, in 1893; it had 5ft 3in driving wheels, two external 18 × 24in cylinders, and weighed 83 short tons in working order. Others were made by the American Locomotive Works of Schenectady for the Boston & Albany Railroad – part of the New York Central Lines – about 1915. These also had 5ft 3in wheels, but the cylinders measured 20 × 24in and weight had risen to 114.5 short tons.

2-6-6T *Breckenridge* was a typical Mason Fairlie or Mason Bogie, made for the Denver, South Park & Pacific Railroad in 1879.

ooOOO

4-6-0, or 2C

Alternative European classifications: 230 or 3/5.
Popular name: Ten Wheeler.

The first examples were used to draw heavy passenger trains up severe gradients. In later days, however, a distinction was made between small-wheel designs intended to haul freight, medium wheeled designs for mixed traffic, and large-wheeled engines for passenger duties. The Ten-Wheeler

A de Glehn/du Bousquet-type compound 4-6-0 for the Russian Ryazan-Ural railways. No. 126 was built in the Putilov factory in 1910, one of 56 U locomotives (there were also six U^{U}, with superheaters). Designed by Mikhail Gololobov, they had two 37 × 58cm (14½ × 22⅞in) high-pressure cylinders externally and two 41 × 58cm (16⅛ × 22⅞in) low-pressure cylinders between the frames; boiler pressure was 14at (206psi) and the driving wheels measured 1.73m (5ft 8in). The last representative of the class survived until 1952.

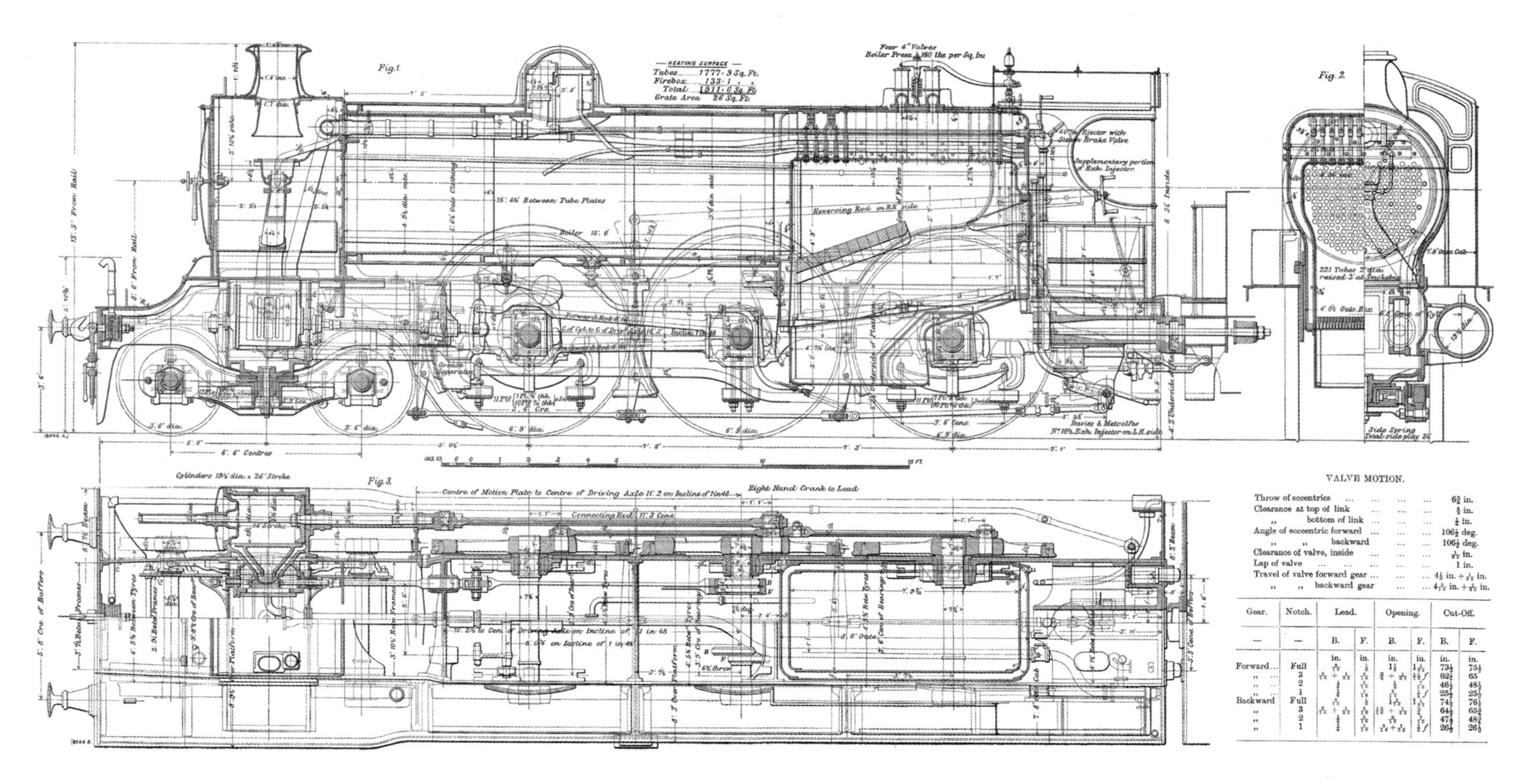

has proved the most versatile of all medium-power categories, though soon displaced in the USA by larger locomotives. The addition of a trailing carrying axle (4-6-2) enabled a larger firebox to be fitted, which was often important if the quality of the firing coal or anthracite was poor.

The earliest 4-6-0, by a narrow margin, was *Chesapeake*, ordered from the Norris Brothers in 1846 by the Philadelphia & Reading Railroad – but possibly designed by John Brandt of the Erie Railroad – and delivered in March 1847. The first to be built in Britain was sent to the Copiapo Extension Railway, in Chile, in 1860. Built by R. & W. Hawthorn of Newcastle upon Tyne, it had 4ft-diameter coupled wheels and was driven

A drawing of a Great Central Railway 195-Class 4-6-0, designed by John G. Robinson. Introduced in 1903, these locomotives had 6ft 9in driving wheels, two 19½ × 26in cylinders and a boiler pressure of 180psi. They weighed 67 tons 13cwt without tender. (*Engineering*, 1903)

by two 16 × 24in outside cylinders. Three coupled axles kept loading below 7 tons.

The first 4-6-0 to run in Britain was Highland Railways *No. 103*, designed by David Jones (1834–1906) and built by Neilson & Company of Glasgow in 1894. No. 103 had two external 20 × 26in cylinders, 175psi boiler pressure and 5ft 3½in coupled wheels; total weight was 94 tons 7cwt. The last British 4-6-0s were Standard Class 5MT made in Derby in 1957, with two 19 × 28in cylinders, 6ft 2in wheels and boilers pressed to 225psi. They weighed 76 tons without tenders.

A group of typically British 4-6-0 locomotives, shown here in model form. Top to bottom: SR No. 771, *Sir Sagramore* of the N15 or King Arthur Class (built in 1925), without smoke deflectors; LNER B17 No. 2846 *Gilwell Park* (1935), shown in British Railways livery as 61646; LMS Black Five No. 4694 (1943) in British Railways livery as 44694; and GWR No. 5075 *Wellington* of the Castle Class (1939) in Brunswick Green. The four locomotives are shown to approximately the same scale, but note how the tenders vary. This was usually due to restrictions on turntable size, particularly on the Eastern Region of the LNER. (Hornby Railways)

ooOOOo

4-6-2, or 2C1

Alternative European classifications: 231 or 3/6.
Popular name: Pacific.

The name derived either from 'trans-Pacific' shipment to New Zealand, or from Missouri Pacific Railroad engines made in 1902 by the Brooks Locomotive Works of Dunkirk. Originally nothing more than an extended Ten-Wheeler with a narrow firebox, the perfected wide firebox Pacific was ideally suited to fast passenger traffic. In the USA, however, even the wide firebox types were replaced rapidly first by the 4-6-4 (Hudson) and then by the 4-8-4 (Niagara).

The first 4-6-2 was designed by George Strong. Built by the Vulcan Iron Works, it entered service on the Lehigh Valley Railroad in 1886. A better design with a wide firebox was the New Zealand Railways 3ft 6in-gauge Q Class, built by the Baldwin Locomotive Works of Philadelphia in 1901–03.

One of the original type of Pacific locomotive, built in 1889 by the Schenectady Locomotive Works for the Chicago, Minneapolis & St Paul Railroad. Note the position of the trailing wheel, which was added to correct weight distribution instead of supporting an extended firebox.

Miniature Pacific *Hurricane* of the 15in-gauge Romney, Hythe & Dymchurch Railway was built by Davey Paxman & Co. Ltd of Colchester in 1927 … and is still going strong, nearly ninety years later.

The Belgian Type I Pacific or 4-6-2 express passenger locomotive of 1934. Note the semi-streamlined appearance, and particularly how the tender sheeting is faired into the profile of the cab. There were four 16½ × 28½in cylinders, boiler pressure was about 265psi, overall length was 80ft and the engine and tender weighed 206 tonnes.

Grand Trunk Western Pacific No. 5834 of the K-4-b-Class was built by Baldwin in 1929. The locomotive had 6ft 1in driving wheels, two 25 × 28in cylinders and a boiler pressure of 215psi; the engine alone weighed about 150 short tons. (Dr Richard Leonard; photographed by David Leonard at Detroit in August 1958)

City of Birmingham was built in Crewe Works in 1939 as one of the Princess Coronation Class of the LMS. Streamlined until April 1946, the locomotive was withdrawn in September 1964 as British Railways No. 46235. The highly successful design was credited to William Stanier, who had trained on the GWR.

The first European engine was compound *No. 4501* of the Chemins de Fer Paris–Orléans, delivered in 1907. The cylinders measured 39 × 65cm (15⅜ × 25⅝in, high pressure) and 64 × 65cm (25¼ × 25⅝in, low pressure), coupled wheels were 1.85m (6ft 1in) and boiler pressure was 16at (235psi). Weight in running order was 91 tonnes (engine only).

The first British 4-6-2 was No. 111 *The Great Bear* of the Great Western Railway, designed by George Churchward (1857–1933) and completed in Swindon Works in January 1908. No. 111 had four 15 × 26in cylinders, 6ft 8½in coupled wheels, and a boiler pressure of 225psi; total weight was 142 tons 15cwt.

The most common British 4-6-2 was the West Country/Battle of Britain Class, most of the 110 being built in the Southern Railway's Brighton works in 1945–51. They had 6ft 2in wheels, 280psi boilers and weighed 128 tons 12cwt. Three 16¼ × 24in cylinders were driven by Bulleid chain drive gear, though many of the group were converted subsequently to use Walschaerts' gear, with a marked loss of individuality.

ooOOOoo

4-6-4, or 2C2

Alternative European classifications: 232 or 3/7.
Popular name: Baltic or Hudson (see notes).
The earliest 4-6-4 locomotives were a pair designed by Gaston du Bousquet (1839–1910) for the Chemins de Fer du Nord and built in France in 1910. Comparable engines were never common in Britain, the arrangement being confined largely to tank engines – the first being built for the London, Tilbury & Southend Railway in 1912, to the design of Robert Whitelegg. The only British Baltic-type tender engine is usually considered to have been the high-pressure Hush Hush (later Class W1) built by the London & North Eastern Railway in 1929 to the design of Nigel Gresley, though now often considered to be '4-6-2-2'.

The Milwaukee Road ordered the first locomotives of this notation in 1925, but bankruptcy meant that the order given in 1928 by the Chicago, Milwaukee, St Paul & Pacific Railroad was not fulfilled by

The imposing 4-6-4T designed by George Hughes of the Lancashire & Yorkshire Railway was built for the LMS after the 1923 grouping. Intended for high-speed suburban passenger traffic, it had 6ft 3in driving wheels, four 16½ × 26in cylinders and a boiler pressure of 180psi. The engine weighed 99 tons 19cwt in working order.

Class H1b 4-6-4 No. 2816 was supplied to the Canadian Pacific Railway by Montreal Locomotive Works, one of ten dating from 1930. There were two external 22 × 30in cylinders, 6ft 3in driving wheels, a boiler pressure of 275psi and a locomotive weight of 159 tons. (Ian McGregor, Canadian Museum of Making)

Baldwin until 1930. Consequently, the first 4-6-4s to run in the USA were built for the New York Central Railroad in 1927, by the American Locomotive Company of Schenectady. The generic name was conferred on them simply because the NYC tracks ran through the valley of the Hudson River. Hudsons soon became popular in the USA, displacing most of the Pacific type (4-6-2) engines from high-speed passenger trains. Some of these were themselves subsequently replaced by 4-8-4s when loads became even greater.

The New York, New Haven & Hartford Railroad called its I-5 Class 4-6-4 locomotives 'Shore Line' or 'Shoreline'. The term 'Royal Hudson' used on the Canadian Pacific Railway was conferred – supposedly only on the streamlined machines – during the tour of Canada by King George VI and Queen Elizabeth in 1939.

Among the last locomotives to be designed by the Nord railway before it was assimilated in 1938 with the state-owned SNCF, 4-6-4 232U1 was not built until 1948. The work of Marc de Caso, 232U1 and its earlier near sisters proved to be capable of indicating in excess of 4,500hp – the most powerful of all French four-cylinder compounds except for the experimental 'one-off' Chapelon 242A1.

Four Driving Axles

OOOO

0-8-0, or D

Alternative European classifications: 040 or 4/4.

Engines in this category were confined to slow freight traffic, owing to absence of guidance for the coupled wheels. They were usually replaced by 2-8-0 and 2-10-0 patterns, being relegated, in North America at least, to the status of switchers and 'yard goats'. The first engine of this type is said to have been the vertical boiler 'grasshopper' *Buffalo*, built in 1841 by Ross Winans of Philadelphia, for service on the Baltimore & Ohio Railroad, but rebuilt in a more conventional 'mud digger' form in 1844. The earliest British equivalents were two tank engines built by the Avonside Engine Company, Bristol, for the Vale of Neath Railway in 1864. They had two

A typical Winans 0-8-0 'mud digger'.

18½ × 24in cylinders inside the frames, 4ft 6in coupled wheels and weighed about 53 tons in running order.

The first 0-8-0 British tender engines were two outside cylinder freighters, purchased by the Barry Railway from Sharp, Stewart & Company of Manchester in 1889. They had been completed three years earlier for the Swedish & Norwegian Railway but had never been delivered. US-type 0-8-0 switchers (a few of which were converted from 2-8-0 simply by removing the pony truck) could be large and powerful. Alton & Southern Railroad No. 12, for example, had three 22 × 28in cylinders, 4ft 9in driving wheels and a boiler pressure of 200psi, giving a calculated tractive effort of 60,600lb. The locomotive weighed 121 short tons and could move surprisingly heavy loads, though only at slow speed.

OO=OO

0-4=4-0, or BB

Alternative European classifications: 0220, 4/4.

The first Mallet-type locomotive, an 0-4-4-0T, was proposed in 1884 but not built for three years by the Decauville company. The first engines were intended for narrow-gauge track, including six supplied for the 60cm 'inner circle' lines constructed to serve the Paris Exposition Universelle of 1889. The first 0-4-4-0 tender engine was built by Société Alsacienne de Constructions Mécaniques in 1894. Supplied to the Prussian state railways, the Königlich Preussische Staats-Eisenbahnen, these locomotives weighed 55.1 tons; tenders added another 39.4 tons. Power was provided by two high-pressure (15.9 × 23in) and two low-pressure cylinders (23.6 × 23in); the diameter of the driving wheels measured 50in. Boiler pressure was 171psi.

OO+OO

0-4-0+0-4-0, or B+B

Alternative European classifications: 020+020 or 4/4.

The earliest locomotives of this type, apparently excepting a geared drive machine proposed (but possibly not built) in the late 1820s by Matthew Murray, were *Wiener-Neustadt* and *Seraing*, built by Günther of Vienna-Neustadt and Cockerill of Seraing for the Semmering trials of 1851. *Seraing* is often regarded as the prototype of the Fairlie system, with two pivoting power bogies, whereas *Wiener-Neustadt* presaged Meyer-type articulation with only one pivoting bogie and a radial slide. Both engines were quite large for their day, *Seraing* weighing 55.1 tons and *Wiener-Neustadt* about 63 tons; boiler pressures were 106psi and 125psi respectively.

The first Garratts were also of 0-4-0+0-4-0 pattern, a pair of 2ft-gauge compounds being supplied by Beyer, Peacock & Co. Ltd to Tasmania in 1909 (pictured on page 69). The two high-pressure cylinders were placed at the inboard end of the rear unit, with the low-pressure cylinders at the inboard end of the front unit.

The use of compounding on Garratts was comparatively unusual, and, in addition, later engines almost always had the cylinders at the outboard ends of the power units to keep them away from the heat of the firebox.

The last Garratt of this notation was supplied in 1937 by Beyer, Peacock & Co. Ltd to Baddesley Colliery.

OOOOo

0-8-2, or D1

Alternative European classifications: 041 or 4/5.
Engines in this category were confined to slow freight traffic, owing to the absence of guidance for the coupled wheels. Though widely applied to tank-engine equivalents of 0-8-0 tender engines, a few tender examples were made with wide or unusually long fireboxes. In addition, a few old US Mikados (2-8-2) were converted to 0-8-2 for use as switchers.

The first 0-8-2T to run in Britain was made in 1896 by Sharp, Stewart & Co. Ltd of Glasgow for the Barry Railway. The locomotive was fitted with 20 × 26in outside cylinders, had 4ft 3in coupled wheels, and its boiler was pressed to 150psi. Weight in running order amounted to 74 tons 16cwt. The engines made in 1908 in the Horwich works of the Lancashire & Yorkshire Railway, to the designs of George Hughes, had two large inside cylinders – 21½in bore, 26in stroke. Boiler pressure was 180psi and the diameter of the coupled wheels was 4ft 6in. Weight in running order approached 84 tons.

LNWR 0-8-2 tank engine No.1548, Crewe works number 5042, dates from December 1911. Designed by Charles Bowen Cooke, it had 4ft 3in driving wheels and two inside cylinders measuring 20½ × 24in. The engine was withdrawn in December 1948 as LMS No. 7888.

OOOOoo

0-8-4, or D2

Alternative European classifications: 042 or 4/6.
Engines of this type were generally tank patterns with bunkers large enough to require a carrying bogie. The first 0-8-4T to run in Britain was a massive banker destined for the Wath marshalling yard on the Great Central Railway between Mexborough and Barnsley. Designed by John Robinson and made in Gorton by Beyer, Peacock & Company, the engine had three 18 × 26in cylinders, 4ft 8in-diameter coupled wheels, and a boiler pressure of 200psi. It weighed 96 tons 14cwt in running order. A booster unit was added to the bogie for a time, but proved to be an unnecessary complication and was soon discarded. Another 0-8-4T banker was made for the London & North Western Railway in 1922. Built in Crewe works to the designs of H.P.M. Beames, it had two 20½ × 24in inside cylinders, 4ft 5½in coupled wheels, and a boiler pressure of 185psi. Weight in running order was about 88 tons.

oOOOO

2-8-0, or 1D

Alternative European classifications: 140 or 4/5.
Popular name: Consolidation.
The original 2-8-0 was a variant of the 4-6-0 for heavy freight work, the substitution of a truck for a bogie and a reduction in the size of the coupled wheels allowing a fourth driven axle to be incorporated within the same overall length. The *Consolidation* was an excellent medium-power freight engine, but was often replaced in the USA and Europe (though not in Britain) by the wide firebox 2-8-2.

The first 2-8-0, with a rigid frame, may have been a conversion of the Pennsylvania Railroad 0-8-0 *Bedford* undertaken in 1864–65 to the designs of John Laird. Designed by Alexander Mitchell and built by the Baldwin Locomotive Works of Philadelphia in 1866, with a leading truck, *Consolidation* celebrated the amalgamation of the Lehigh & Mahony and Lehigh Valley railroads.

Though 2-8-0s were being made by Dübs & Company of Glasgow as early as 1884 – for the Brazilian Paulista railway – the first to run in Britain did not appear until the early 1900s, replacing the slow-speed 0-8-0 on perishable or high-speed freight. The earliest British service representative was *No. 2800* of the Great Western Railway, designed by George Churchward and built in Swindon works in 1903. The engine had two 18 × 30in outside cylinders, 4ft 7½in-diameter coupled wheels and weighed 68 tons 7cwt (engine only). The boiler pressure was 225psi.

The first 2-8-0 to run on the Great Western Railway was No. 97, later No. 2800, outshopped from Swindon in 1903 to the design of George Churchward. Seen as part of a standardisation programme, No. 2803 of 1905, pictured in photographic grey, was among the first of hundreds of essentially similar locomotives.

Except for the post-war 'O1' engines of the LNER, which were rebuilds of older machines, the last 2-8-0s to be made in quantity were the 'Austerity' class. The first engine was completed by the North British Locomotive Co. Ltd, Glasgow, in January 1943, total production amounted to more than 1,200. They had two 19 × 30in cylinders, 4ft 8½in coupled wheels and weighed 125 tons 15cwt with tender. Boiler pressure was 225psi.

The original *Consolidation* 2-8-0 was built in 1866 by the Baldwin Locomotive Works of Philadelphia for the Lehigh Railroad.

A 2-8-0 *Consolidation*, No. 1799, made by Baldwin for the Rock Island Line. Note that this locomotive is fitted with Walschaerts' gear, uncommonly seen in the USA in the twentieth century.

oOO=OO

2-4=4-0, or 1BB

Alternative European classifications: 1220, 4/5.
The first engines of this type were made *c.* 1897 by the Kolomna locomotive works, for the Trans-Siberian Railway. They weighed 56 tons without their tenders and were driven by a pair of high-pressure cylinders (18.7 × 26in) exhausting to a pair of low-pressure units (28 × 26in). These Mallets served successfully, often in sub-zero conditions, for many years.

oOOOOo

2-8-2, or 1D1

Alternative European classifications: 141 or 4/6.
Popular names: Calumet or Mikado (see notes).
The original 2-8-2 was a variant of the 2-8-0 freight engine, the additional carrying axle allowing a wide firebox to be fitted. Some of the largest wheeled versions, however, were conceived for arduous high-speed passenger work.

The first 2-8-2 was made for the Calumet & Hecla Mining Company in the 1880s, probably by the Baldwin Locomotive Works of Philadelphia. Baldwin subsequently supplied comparable engines to Japan in 1897, a new popular name deriving from the title of the Japanese emperor. Customarily known as the *Mikado* (abbreviated to Mike), a few locomotives of this notation, the MAC-2-61 Class used by the Minneapolis & St Louis Railroad, were known as *MacArthurs*. Originally made by Alco in 1916, they were rebuilt in the M&SLRR Cedar Lake shops in 1942 as part of the war effort.

Grand Trunk Western Class S-3-c No. 3748, built by Alco in 1924, had 5ft 3in driving wheels, two 26 × 30in cylinders, a boiler pressure of 200psi and weighed 154 short tons (locomotive only). (Dr Richard Leonard; photographed by David Leonard at Detroit in September 1958)

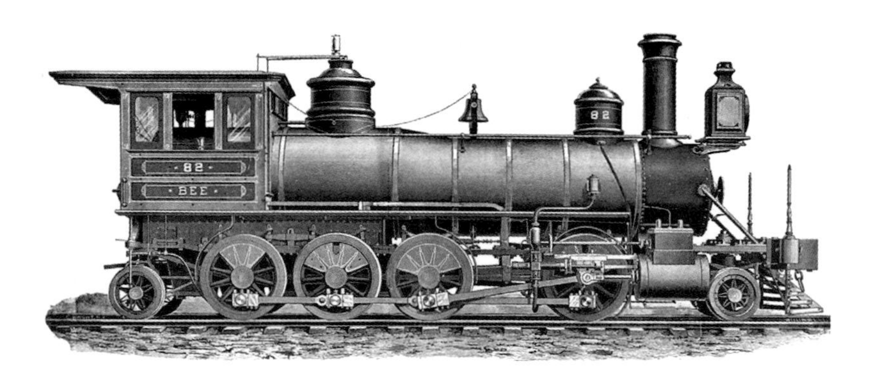

The strange wheel spacing of 2-8-2 *Bee* suggests problems with weight distribution, but the engine had been built as a 2-10-0 in 1867 and only converted to 2-8-2 notation in 1883.

Cock o' the North was the first 2-8-2 to be built for service on a British railway. Designed by H.N. Gresley for service on the east coast of Scotland, the P2 *Mikados* were the most powerful express passenger locomotives to run in Britain. They were eventually rebuilt by Edward Thompson in 4-6-2 form, in 1944, which completely spoiled their appearance.

The first Mikados to run in Britain were P1 freight engines of the London & North Eastern Railway, designed by Nigel Gresley and built in Doncaster in 1925. *Cock o' the North* (P2 Class), designed by Gresley for fast passenger traffic on routes operated by the London & North Eastern Railway on the east coast of Scotland, was the most powerful locomotive to run in Britain except for the Garratts and the 0-10-0 LMS *Lickey Banker*. Completed in Doncaster works in May 1934, No. 2001 *Cock o' the North* had cam-operated Lentz poppet valves, three 21 × 26in cylinders, 6ft 2in driving wheels and a boiler pressure of 220psi. Weight in running order was 165 tons 11cwt. The Lentz valves were replaced by Walschaerts' gear when *Cock o' the North* was rebuilt in 1939 to resemble the streamlined A4 Class (aesthetically a retrograde step in the view of many observers!).

Among the most successful 2-8-2s were the powerful and smooth-running SNCF-Class 141P compounds, introduced in 1941, which hauled heavy freight and fast passenger trains until the 1970s. The cylinders measured 41 × 70cm (high pressure) and 64 × 70cm (low pressure), boiler pressure being 20at (294psi). Coupled wheels measured 1.65m and the engines weighed 112.5 tonnes without their tenders.

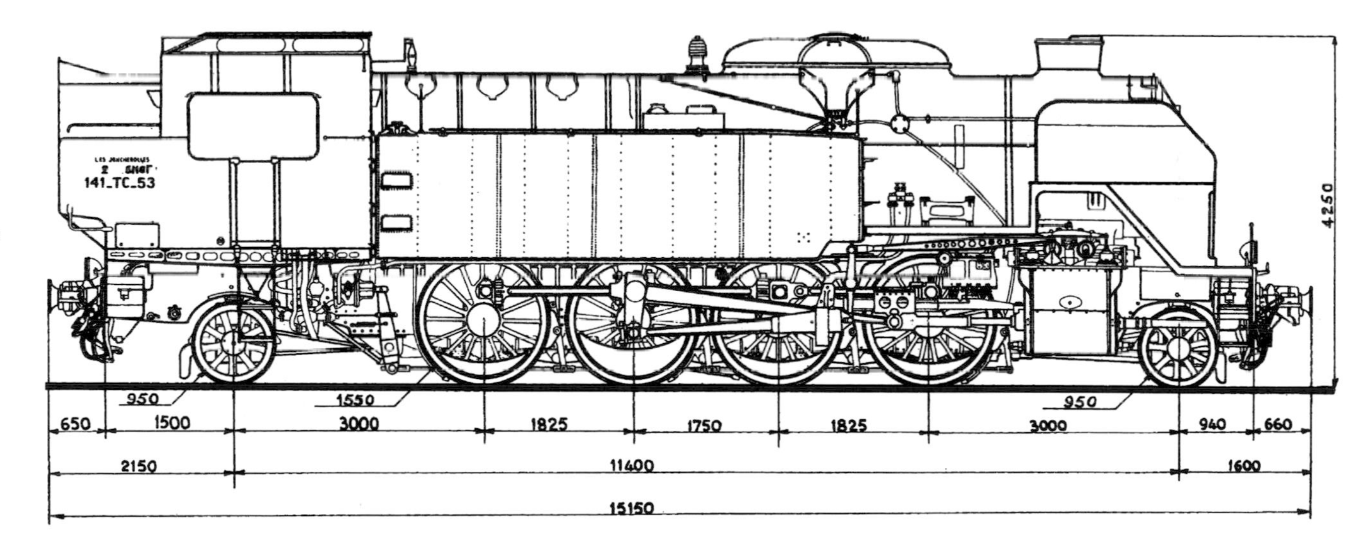

A line drawing of SNCF 2-8-2T No. 141 TC 53 (the original was made by Schneider in 1934), showing the many refinements to be expected in a French design of the late 1930s. The work of Marc de Caso, they had Cossart valve gear and 5ft 1in coupled wheels; weight was about 121 tons.

oOO=OOo

2-4=4-2, or 1BB1

Alternative European classifications: 1221, 4/6.
Mallets of this type were rarely encountered, though some were supplied to Russian railways prior to 1914.

oOO+OOo

2-4-0+0-4-2, or 1B+B1

Alternative European classifications: 120+021 or 4/6.
The first Garratts of this notation were supplied by Beyer, Peacock & Co. Ltd to the 5ft 3in-gauge São Paulo Railway in Brazil. Dating from 1915, they had four 16 × 24in cylinders, 5ft-diameter driving wheels and weighed 87 tons 9½cwt. The last representative was a solitary locomotive sent by Beyer Peacock to Ceylon in 1930.

oOOo+oOOo

2-4-2+2-4-2, or 1B1+1B1

Alternative European classifications: 121+121 or 4/8.
The only Garratts of this notation were four supplied by Beyer, Peacock & Co. Ltd to the metre-gauge Leopoldinha Railway in Brazil in 1943.

oOOOOoo

2-8-4, or 1D2

Alternative European classifications: 142 or 4/7.
Popular name: Berkshire (see notes).
This was basically either a 2-8-2 with an elongated firebox or a 2-8-2T with an extended bunker. No locomotive of this layout has ever run in Britain, but engines were made by NBL for South Africa and elsewhere.

The original 2-8-4 was developed by the Lima Locomotive Works for the Boston & Albany Railroad, the first example being supplied early in 1925. 'Super Power' designs incorporating a very large firebox, necessitating the trailing bogie, they were intended to work over the Berkshire Hills in Massachusetts. Though widely known as 'Berkshire' on most railroads, the Class K-4 2-8-4s of the Chesapeake & Ohio Railroad (Alco, 1947) were listed as 'Kanawha'. Austria, Germany (East and West) and the USSR used 2-8-4s in quantity, and sixty-six French-built Chapelon engines went to Brazil in 1951–52.

The Soviet *Iosif Stalin* or IS 2-8-4, designed by a team led by Nikolai Suschkin of the Central Locomotive Project Group, was introduced to traffic in 1932; 649 of them were made in Kolomna and Voroshilovgrad (Lugansk) until the German invasion of Russia in June 1941 stopped work. They had 1.85m (6ft 0⅞in) driving wheels, two 67 × 77cm (26⅜ × 30⅓in) cylinders and a boiler pressure of about 15at (221psi). The locomotive weighed about 133 tonnes without its tender. A few streamliners were made in the late 1930s and the last standard IS remaining in service was not withdrawn until 1972.

oOOOOooo

2-8-6, or 1D3

Alternative European classifications: 143 or 4/8.

The 2-8-6 was basically a 2-8-4 with an unusually large firebox, intended to burn low-grade coal. The prototype 2-8-6 was apparently developed by the Lima Locomotive Works, but it is not known whether any others were ever made. The size of the firebox necessitated an unusual three-axle bogie, shared with some of the company's articulated Mallets.

oOOOO+ooo

2-8-6, or 1D3

Alternative European classifications: 143, 4/4.

Made only for the Denver, South Park & Pacific Railroad, about 1890, these large Mason Fairlie locomotives had a leading truck to improve high-speed running and large fuel bunkers extending backward above the carrying bogie.

ooOOOO

4-8-0, or 2D

Alternative European classifications: 240 or 4/6.

Popular name: Twelve-Wheeler or Mastodon.

The earliest of these were modifications of the Ten-Wheeler, generally in an attempt to reduce axle loading. The first was built for the Baltimore & Ohio Railroad by Ross Winans in 1854. Delivered in 1855, *Centipede* was originally a cab-forward design; in 1865, however, a conversion to Camelback form was undertaken.

Among the first locomotives of this type were *Champion* of the Lehigh Valley Railroad and *No. 229* of the Central Pacific Railroad, completed in Weatherly and Sacramento respectively in 1882, but the notation

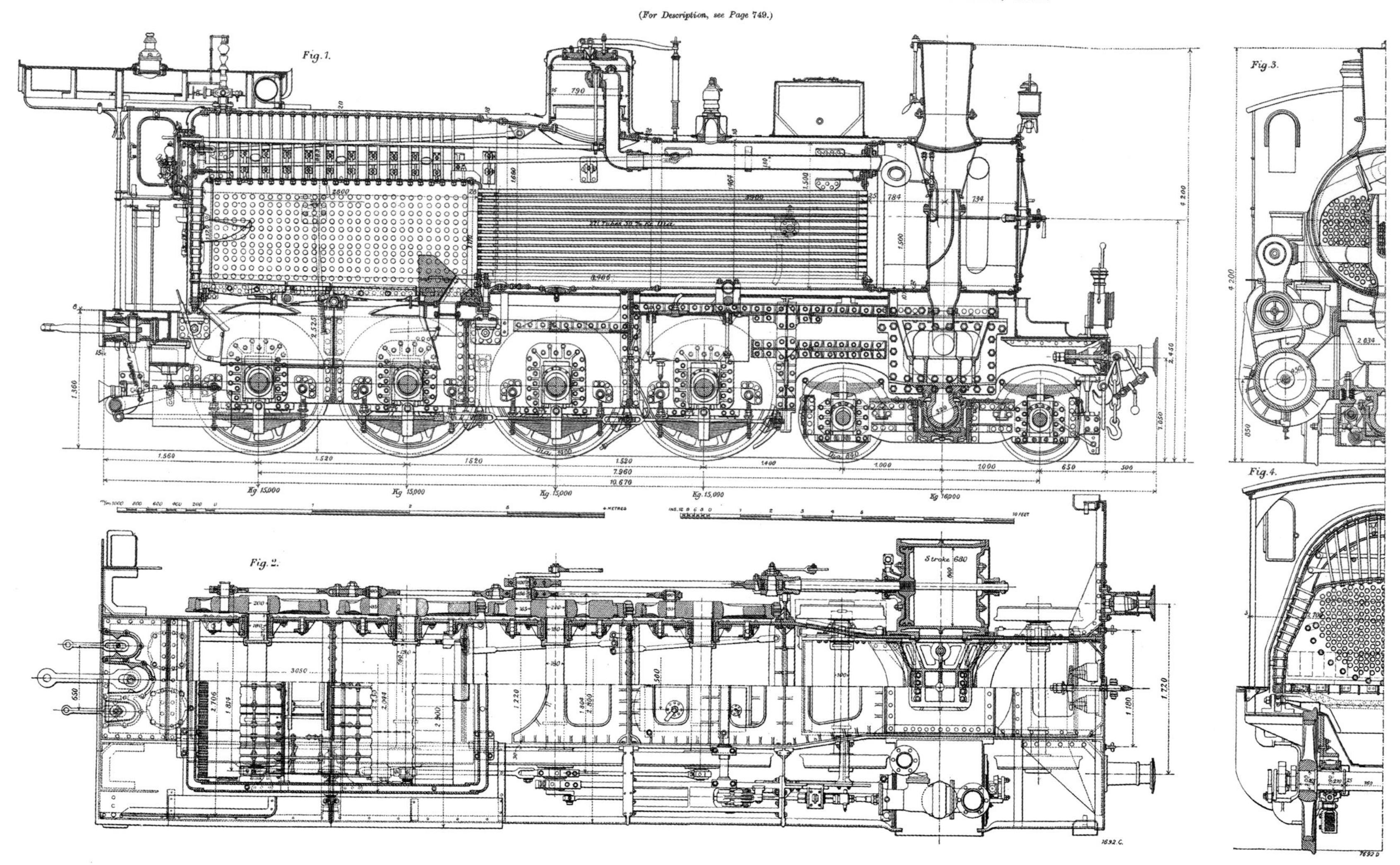

A two-cylinder compound 4-8-0 made by Ernesto Breda of Milan for the Mediterranean Railway Company.

was not popular in North America. However, the Chicago, Indianapolis & Louisiana Railroad two-cylinder Class E-1a, delivered from the Brooks Locomotive Works in 1898–1900, lasted in service until 1949. Built with Stephenson valve gear, slide valves and a Belpaire boiler, the engines were rebuilt in 1922 with Walschaerts' gear, piston valves and an American-style round-top boiler.

The 4-8-0 locomotives of 33. Klasse of the Austrian federal railways were intended for express passenger traffic on comparatively lightly laid track, the fourth coupled axle allowing sufficient adhesion. The driving wheels measured 1.74m (5ft 8½in), there were two 60 × 72cm (23.6 × 28.4in) cylinders driven by Lentz valve gear, and the boiler pressure was about 15at (221psi). Overall length was 20.7m (67ft 11in) and the engine weighed 85.2 tonnes in working order. Forty were built in 1923–28, the last of them surviving until 1968.

The Twelve-Wheeler enjoyed a renaissance in the 1930s, particularly in France and on broad-gauge lines in South America.

Many 4-8-0s were built in Britain for colonial service, but the only standard-gauge engines of the type to run in Britain were tank engines. The first of fifteen 4-8-0Ts was built in 1909 in the Gateshead works of the North Eastern Railway to the design of Wilson Worsdell. Intended for use in the Erimus marshalling yard, near Middlesbrough, the engine had three 18 × 26in cylinders, 4ft 7¼in coupled wheels, and a weight of 84 tons 13cwt. Boiler pressure was 175psi. The G16-Class 4-8-0T was introduced on the London & South Western Railway in 1921, to shunt in Feltham yard. Designed by David Urie, the four locomotives had two 22 × 28in outside cylinders and 5ft 1in coupled wheels. Boiler pressure was 180psi and they weighed 72 tons 18cwt in working order.

The Londonderry & Lough Swilly Railway owned two 4-8-0 tender engines, built by Hudswell Clarke & Co. Ltd in 1905, but these ran on 3ft-gauge track.

Among the most numerous of all 4-8-0s was the 424 Class of the Hungarian state railways, 365 being made in 1924–56. They had two 60 × 66cm cylinders, 1.61m-diameter coupled wheels and a modest 14at (206psi) boiler pressure. Engine weight was about 143 tonnes.

ooO=OOo

4-2=4-2, or 2AB2

Alternative European classifications: 2121, or 3/6.

A locomotive of this type, apparently a Webb-type compound tank engine, is said to have been supplied by Robert Stephenson & Company of Newcastle upon Tyne to a mine railway in Antofagasta, Chile, in 1884. It is not entirely clear how the drive was arranged, nor how many cylinders were used.

ooOOOOo

4-8-2, or 2D1

Alternative European classifications: 241 or 4/7.
Popular name: Dübs or Mountain.
The earliest of these were modifications of the 4-8-0, generally in an attempt to reduce axle loading, increase the diameter of the driving wheels or allow a larger firebox to be fitted. Most machines were known as 'Mountains', though the New York Central, which ran over essentially flat terrain, called its locomotives 'Mohawks'.

The earliest 4-8-2 locomotives were made in Britain for colonial service, the name of their manufacturer – Dübs & Company of Glasgow – providing the original popular name. None has ever run on standard-gauge railways in Britain, except for modifications of the 4-8-0 North Eastern Railway tank engines mentioned earlier. However, the 15in-gauge Romney, Hythe & Dymchurch Railway still runs No. 5 *Hercules* and No. 6 *Samson*, built by Davey, Paxman & Co. Ltd in 1926.

The first US engine of this type appeared on the 'mountain' section of the Chesapeake & Ohio Railroad in 1911, but the near-universal adoption of wide fireboxes after the end of the First World War favoured the 4-8-4.

Rutland Railroad L-1 Class 4-8-2 No. 90 was delivered from Alco in June 1946 but only lasted in service until 1955. The locomotive had 6ft 1in driving wheels, two 26 × 30in cylinders, a boiler pressure of 230psi and weighed 174 short tons without tender. (From a picture postcard; courtesy of Dr Richard Leonard)

ooOOOOoo

4-8-4, or 2D2

Alternative European classifications: 242 or 4/8.
Popular name: Northern, but see notes.
This was an enlargement of the 4-8-2, with a wide firebox. The 4-8-4 proved an ideal combination of power and weight, and was extremely

Union Pacific 4-8-4 No. 814 of FEF-1 Class, seen at Grand Island, Nebraska, in August 1957, was built by Alco in 1937. The diameter of the driving wheels was 6ft 5in, the cylinders were 24½ × 32in and boiler pressure was 300psi; this gave a tractive effort of 63,610lb. Overall length was 98ft 5in; weight was 415 short tons. (Dr Richard Leonard; photographed by David Leonard)

About 251 examples of the Soviet P-36 4-8-4 were made in the Kolomna locomotive works in 1952–56. They had 1.85m (6ft 0.8in) driving wheels, two 57.5 × 80cm (22.6 × 31.5in) cylinders and a boiler pressure of 15at (221psi). The last to be made, P36-0251, was also the final Soviet steam locomotive, as a decision to introduce more modern forms of traction had been taken.

Chapelon's masterpiece: 4-8-4 242A1, scrapped by the SNCF before it could eclipse the first generation of electric and diesel locomotives.

successful in North America hauling heavyweight freight and high-speed passenger trains at 100mph or more. This class was best known as the *Northern*, the first Baldwin-made A-1 being delivered to the Northern Pacific Railroad in 1926. The sobriquet was accepted by many US railroads, particularly those running in the northern Midwest. They included the Canadian National; the Chicago, Burlington & Quincy; the Chicago, Milwaukee, St Paul & Pacific; the Chicago & North Western; the Chicago, Rock Island & Pacific and the Wisconsin Central.

The name spread eastward to railroads such as the Atlantic Coast Line and even the Delaware & Hudson. However, an assortment of alternatives graced what became one of the most widely distributed of all the wheel arrangements tried in the USA. The Nashville, Chattanooga & St Louis designated their locomotives 'Dixie', beginning with a batch delivered by Alco in 1930 and ending with the J3-57 Class of 1943; the Chesapeake & Ohio Railroad named its Lima-made J-3-A machines 'Greenbriar'; The Delaware, Lackawanna & Western Railroad labelled its 1932 vintage Alco-built Q3 locomotives 'Pocono'; and the Lehigh Valley Railroad called its T-3-B 4-8-4s delivered by Alco in 1943 'Wyoming'.

Many more were built for service in the USA, and other 4-8-4s were to be seen in Europe and southern Africa. Typical of these large machines was the Southern Pacific-Class GS 4, introduced in 1941, which had two 25½ × 32in outside cylinders and 80in coupled wheels; boiler pressure was 280psi. GS 4 locomotives weighed 424 short tons with their tenders.

No 4-8-4 tender engine has ever run on British tracks, though many have been built in Britain for service overseas – e.g. the KE7-Class engines supplied by the Vulcan Foundry of Newton le Willows to the Chinese railway service in the 1930s. These had two outside cylinders measuring 20¼ × 29½in, boilers pressed to 220psi and 5ft 9in-diameter coupled wheels. They weighed 196 tons with tender.

Two 4-8-4 tank engines ran on the Burtonport Extension of the Londonderry & Lough Swilly Railway (3ft gauge). Built in 1912 by Hudswell Clarke & Co. Ltd, they weighed 51 tons.

The most efficient example was the solitary SNCF No. 242A1 of 1945, created by the great engineer André Chapelon. This three-cylinder compound weighed only 225 tonnes with its tender, had 1.95m-diameter coupled wheels and could develop more than 4,000hp at the drawbar. Cylinder dimensions were 60 × 72cm (high pressure) and 68 × 76cm (low pressure); boiler pressure was 20.5at (301psi). Unfortunately, the locomotive appeared at a time when modernisers were determined to introduce diesel and electric traction. Consequently, it was given little chance to show its potential.

ooOO÷OOoo

4-4÷4-4, or 2BB2

Alternative European classifications: 222 or 4/8.
Unique to the T1 duplex-drive locomotives of the Pennsylvania Railroad.

The Pennsylvania Railroad 4-4÷4-4 T1 duplex was an extraordinary design with 6ft 8in driving wheels, four 19¾ × 26in cylinders (as built), oscillating-cam poppet valves and 300psi boiler pressure. No. 5527 was supplied by Baldwin in 1946, but the entire class of fifty had been withdrawn by 1954. (Dr Richard Leonard)

ooOOo+oOOoo

4-4-2+2-4-4, or 2B1+1B2

Alternative European classifications: 221+122 or 4/10.
Two compound Garratts of this type were supplied to Tasmania in 1912. Unique among the Beyer, Peacock & Co. patterns, they had four cylinders on each power unit – two high pressure inside the frames and two low pressure outside. The only other representatives of this type were five simple expansion engines supplied in 1927 to the Entre Rios Railway in Brazil and three that were dispatched to to the North Eastern Railway of Argentina in 1930.

oooOO÷OOooo

6-4÷4-6, or 3BB3

Alternative European classifications: 323 or 4/10.
Unique to the solitary S1 duplex-drive locomotives of the Pennsylvania Railroad.

The Pennsylvania Railroad S1 6-4÷4-6 duplex was the precursor of the T1. Sharing somewhat similar characteristics (but larger), the locomotive gave constant trouble and, owing to poor adhesive weight, was prone to excessive slipping.

Five Driving Axles

OOOOO

0-10-0, or E

Alternative European classifications: 050 or 5/5.
Popular name: Decapod (Britain).
This layout, with adhesive weight equal to the total available, was an ideal slow-speed heavy haulier. Though confined in the USA largely to

The LMS 0-10-0 No. 2290 *Decapod* or *Lickey Banker* was designed by James Anderson of the Midland Railway. Built in Derby works in 1919, the unique locomotive weighed 73 tons 13cwt. It had 4ft 7½in wheels and four 16¾ × 28in cylinders with Walschaerts' gear. Although boiler pressure was only 180psi, the tractive effort of 43,300lb was the greatest of all rigid frame engines to run on a British railway.

short distance or shunting work, the 0-10-0 has been extremely popular in continental Europe. About 14,000 were built in Russia and the USSR alone. The US railroads preferred heavy freight engines with trailing trucks or bogies, allowing a wide firebox to be used, whereas exceptionally small coupled wheels (and high ground clearance in Russian examples) allowed the European designs to mount a comparatively narrow firebox of sufficient depth.

Early examples included *Steierdorf*, designed by John Haswell and built for the Imperial Austrian Railway in 1861, which had three axles in the mainframe and two in a sub-frame to the rear, connected by a parallel motion credited to Pius Fink. This allowed the locomotive to take surprisingly tight curves, but the geometry was ineffectual and the connectors wore excessively. Some of the original Engerth locomotives were also of 0-10-0 form, though the connection between the third and fourth axle was achieved by gears (giving the appearance of 0-6-4-0).

Reuben Wells of 1868, an 0-10-0T, was built in 1868 for the Jefferson, Madison & Indianapolis Railroad to take heavy freight trains up a 1-in-18 gradient on Madison Hill. The first British 0–10–0 was a unique tank engine, designed by James Holden, built in the Stratford Works of the Great Eastern Railway in 1902. Three 18½ × 24in cylinders were fitted, the diameter of the coupled wheels was restricted to 4ft 6in, boiler pressure was 200psi and weight totalled 80 tons. The *Decapod* was used successfully to demonstrate that a steam engine could accelerate as rapidly as competing forms of traction, but was notoriously heavy on the track and was rebuilt as an odd-looking 0-8-0 in 1906.

The only British tender engine in this group was the *Lickey Banker*, a four-cylinder example built in the Midland Railway's Derby workshops in 1919 to the designs of James Anderson (the CME, Henry Fowler, was still away on war service). It had four 16¾ × 28in cylinders, 4ft 7½in coupled wheels and a boiler pressure of 180psi. The engine alone weighed 73 tons 13cwt.

OOOOOo

0-10-2, or E1

Alternative European classifications: 051 or 5/6.
Popular name: Union.
This particular layout was confined in the USA largely to short distance or shunting work. The trailing truck allowed an extra large and extra wide firebox to be used. The arrangement seems to have been embodied in a handful of powerful freight engines, such as the S7 Class built in 1936–37 by the Baldwin Locomotive Company for the Union Pacific Railroad. These were intended to transfer iron ore trains from yard to yard in the Pittsburgh area, at very slow speeds. A few European narrow-gauge tank engines of this notation have also been made.

oOOOOO

2-10-0 or 1E

Alternative European classifications: 150 or 5/6.
Popular name: Decapod (USA).
Locomotives of this type were the largest to be made in quantity outside the USA. The addition of a guiding truck enabled greater speeds to be achieved than with the classic 0-10-0. However, even five coupled axles posed problems on curves unless combinations of flangeless wheels and lateral axle movement were used. The US railroads preferred heavy freight engines with trailing trucks or bogies, allowing a wide firebox to be used, whereas exceptionally small coupled wheels allowed the European designs to mount a comparatively narrow box of sufficient depth.

The first engine of this type was designed for the Lehigh Valley Railroad by Alexander Mitchell and, apparently, built by the Baldwin Locomotive

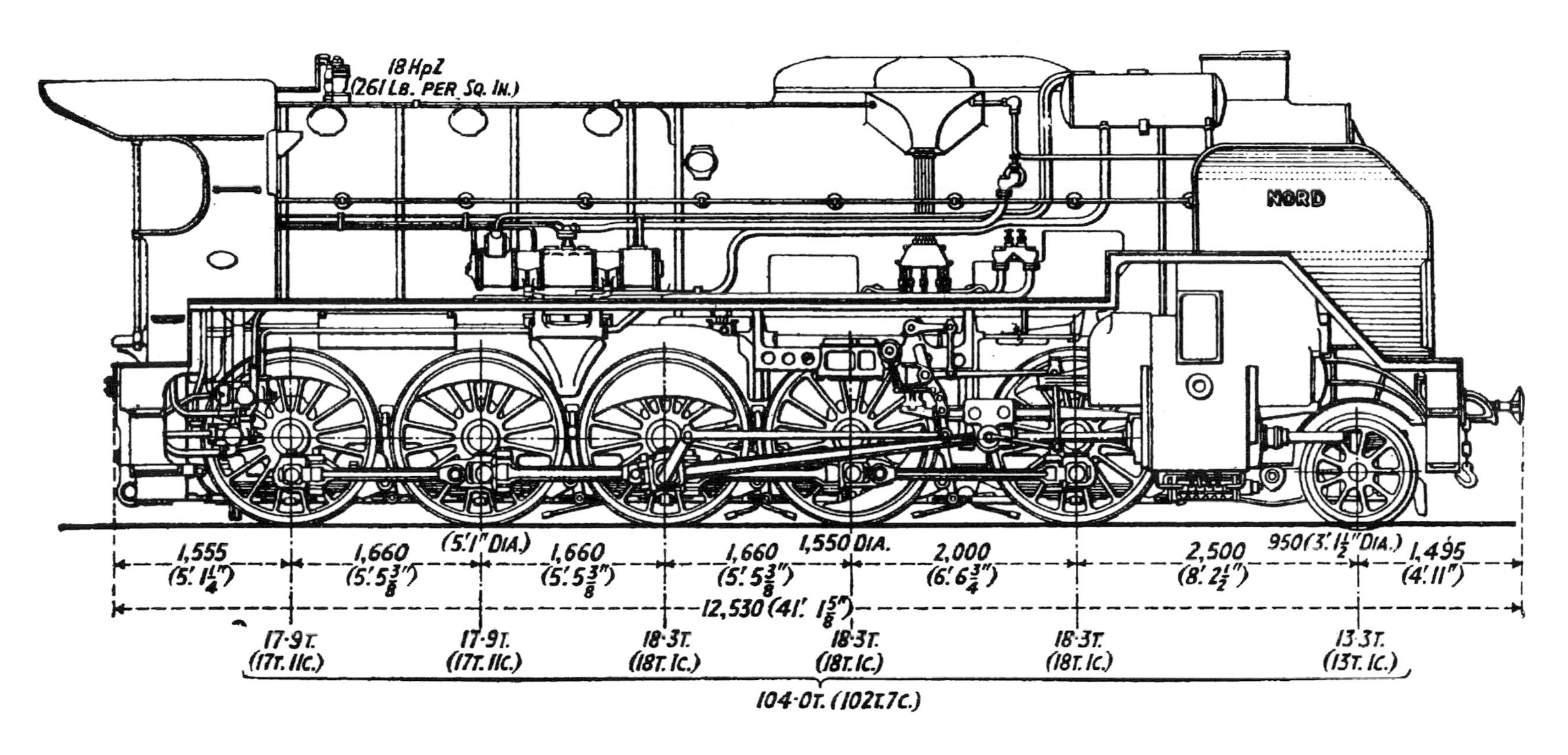

Nord 2-10-0 four-cylinder compound 5.1211, later SNCF 150P Class, had two high-pressure cylinders outside and two low-pressure cylinders inside the frames. The engine weighed 104 tonnes without its tender, and had 1.55m (5ft 1in) driving wheels. Boiler pressure was 18at (261psi). (*The Model Engineer and Practical Electrician*, 1 March 1934)

Works of Philadelphia in 1867. Though large numbers of 2-10-0s were built in the USA during the First World War for tsarist Russia, and many subsequently ran in North America when the 1917 revolution stopped deliveries, the best-known design was the German Class 52 *Kriegslok*. More than 6,000 were built during the Second World War, ensuring that locomotives remained in service throughout Europe into the 1970s. They had two 60 × 66cm outside cylinders, boilers pressed to 16at (235psi), 1.4m coupled wheels, and weighed about 147 tonnes with the finalised four-axle 'tub' tender.

No 2-10-0s were built for service in Britain until the War Department Austerity Class appeared in 1944. However, British Railways built large numbers of the Standard 9F – the last of all, *Evening Star* being completed in Swindon Works in 1960. Two 20 × 28in cylinders, 5ft wheels and boiler pressures of 250psi were standard. Engine weight was 86 tons 14cwt.

oOO=OOO

2-4=6-0, or 2BC

Alternative European classifications: 2230, 5/6.

A few Mallett-type locomotives of this notation were supplied to the Hejaz Railway (1.05m gauge) prior to the First World War. Made by Henschel, they weighed 52.2 tons (engine) and 37.9 tons (tender), had 42in-diameter coupled wheels and a boiler pressed to 171psi. The two high-pressure cylinders had 12.6in bores and a stroke of 23in; the diameter of the low-pressure cylinders was 20in, the stroke-length being common to both sets.

oOOOOOo

2-10-2, or 1E1

Alternative European classifications: 151 or 5/7.
Popular name: Santa Fe.

This layout was confined to the USA and largely to long-distance freight. The trailing truck allowed a large, wide firebox to be used. The original engines were tandem compounds, built by Baldwin of Philadelphia for the Atchison, Topeka & Santa Fe Railroad from 1903 onward. The 3800 Class, supplied by Baldwin in 1919–27, had 5ft 3in driving wheels, two 20 × 32in cylinders and a boiler pressure of 220psi. The locomotive weighed 138 short tons without its tender, and had a tractive effort of about 85,000lb.

The 2850 Class 2-10-2 of the Illinois Central Railroad, built by Lima in the early 1920s and rebuilt in the ICR Paducah shops in 1943, was often known as the 'Central'.

oOOOOOoo

2-10-4, or 1E2

Alternative European classifications: 152 or 5/8.
Popular name: Texas or Selkirk.

Confined largely to long-distance North American freight, this was another of the Lima 'Super Power' designs – a Santa Fe (2-10-2) extended to accept a trailing bogie. This allowed a wide firebox to be fitted. The first example may have been Atchison, Topeka & Santa Fe No. 3829, a 2-10-2 (see preceding entry) that had been given a trailing bogie to investigate if additional support for the firebox was beneficial. The earliest true 2-10-4s, however, were built for the Texas & Pacific Railroad (hence 'Texas') in 1925, though the later T1 or Selkirk Class of the Canadian Pacific Railway was better known. The first of these was delivered from the Montreal Locomotive Works in 1929.

ooOOOOO

4-10-0, or 2E

Alternative European classifications: 250 or 5/7.
Popular name: Mastodon.
This particular wheel arrangement was confined largely to freight engines with wheels small enough to accommodate a narrow firebox between the frames. The remarks made about the renaissance of the 4-8-0 apply equally to the 4-10-0. The earliest of these was *El Gubernator* of the Central Pacific Railroad, built in 1884 by the Baldwin Locomotive Company. The only modern representatives were the three-cylinder Class-11 engines built in 1941 for the Bulgarian State Railways by Henschel & Sohn of Kassel.

ooOOOOOo

4-10-2, or 2E1

Alternative European classifications: 251 or 5/8.
Popular name: Overland (on Union Pacific).
This was a logical development, by Alco, of the 2-10-2 to provide a more powerful three-cylinder locomotive of the same general dimensions. The additional weight of the third cylinder, and the need of space within the frame, was readily solved with a bogie. One locomotive was built as a demonstrator in 1925, but only the Southern Pacific (forty-nine of them in 1925–27) and the Union Pacific (ten in 1926–27) bought 4-10-2s in quantity. Union Pacific SP-1 Class had 5ft 3½ driving wheels, two 25 × 28in low-pressure cylinders mounted outside the frame, one 25 × 32in high-pressure cylinder within the frames and a boiler pressure of 225psi. The locomotives had a tractive effort of 86,589lb and weighed about 223 short tons with their tenders.

ooOO–OOOo

4-4=6-2, or 2BC1

Alternative European classifications: 2231, 5/8.
These interesting engines seem to have been unique to the Atchison, Topeka & Santa Fe Railroad. Dating from *c.* 1910, they were conversions of an Atlantic (front) and a large-wheel Prairie (rear) to make a single high-speed passenger locomotive capable of hauling heavy loads. The engines were distinguished by coupled wheel diameters of 6ft 1in and could run at speeds as high as 70mph. Very few were made, as the conversions were not only wasteful but also poor steamers. Retaining the grate and boiler of the original 2-6-2 restricted steam-raising capacity, even though a feedwater heater was fitted in the extended boiler barrel.

ooOO÷OOOoo

4-4÷6-4, or 2BC2

Alternative European classifications: 252 or 5/9.
The unique Q1 duplex-drive locomotive of the Pennsylvania Railroad was delivered by Baldwin in 1942. The cylinders of the rear axle group were placed by the firebox, where they were exposed to heat and dust. The production version, the Q2 (below), reverted to more conventional form.

ooOOO÷OOoo

4-6÷4-2, or 2CB2

Alternative European classifications: 252 or 5/9.
The largest and heaviest, and possibly also the most powerful rigid frame design ever built, the 26 Q2 duplex-drive locomotives made by Baldwin for the Pennsylvania Railroad in 1944 reverted to a conventional layout with the rear cylinders at the front of the power unit. The tendency to slip violently was supposedly controlled by an anti-slip system, but the Q2 proved to be temperamental in service; all of them had been withdrawn by 1951 and scrapped.

Six Driving Axles

OOOOOO

0-12-0, or F

Alternative European classifications: 060 or 6/6.
This particular wheel arrangement was confined largely to banking and heavy freight engines with coupled wheels small enough to accommodate a narrow firebox between the frames. The earliest representative of this type was 1863-vintage *Pennsylvania*, a Camelback tank engine built in the workshops of the Philadelphia & Reading Railroad to the designs of James Millholland. Intended largely for banking duties, the locomotive's wheelbase proved to be far too long and rigid; consequently, it was rebuilt as an 0-10-0 tender engine in 1870.

The Austrian state railway factory in Floridsdorf, now a suburb of Vienna, built three rack and adhesion 0-12-0T (269-Class) locomotives in 1911–12. These had two 57 × 52cm cylinders, and two auxiliary 42 × 45cm cylinders on the rack engine. Driving wheel diameter was 1.08m, boiler pressure was 12.5at (184psi) and the machines weighed about 88 tonnes in running order.

The only large-scale exploitation of the 0-12-0 layout was the ten Bulgarian state 40-Class two-cylinder compound tank engines built in 1922 by Hanomag–Hannoversche Maschinenfabrik AG. These had cylinder dimensions of 62 × 70cm (high pressure) and 89 × 70cm (low pressure); boiler pressure was 13.6at (200psi). The diameter of the driving wheels was 1.35m and the locomotives weighed 102 tonnes in working order. Powerful and effective, they had lengthy service careers.

The unique 0-12-0T designed by James Millholland for the Pennsylvania & Reading Railroad in 1863.

OOO:OOO

0-6=6-0, or CC

Alternative European classifications: 0330, 6/6.
The first machine of this type seems to have been built for the Chemins de Fer du Nord (France) in about 1860, to the designs of Jules Pétiet. One was shown at the London Exhibition of 1862, together with an essentially similar express passenger engine with a single driving axle at each end of the frame separated by three carrying axles. The Nord locomotives were also remarkable for the curious design of the chimney, which ran back horizontally above the boiler to exhaust directly above the cab. The advent of such a large engine, with all its weight available for adhesion, inspired many engineers to produce similar effects on the basis of established designs. In Britain, Archibald Sturrock produced a series of heavy freight engines for the GNR with an additional three-axle drive unit beneath the tender body.

These ran for a few years, but were disliked heartily; the boiler was not always capable of supplying enough steam to the four cylinders, the footplate became unbearably hot and the lengthy steam pipes were difficult to lag satisfactorily. However, this did not prevent a variety of 'steam tenders' appearing in Europe in the mid-1860s.

OOO=OOO

0-6=6-0, or CC

Alternative European classifications: 0330, 6/6.
The first Mallet of this type, a tank engine, was built as early as 1889 by J.A. Maffei of Munich, for the St Gotthard railway in Switzerland. It weighed a remarkable 85 tons – the largest engine in the world at the time of completion. The first tender engine seems to have been built in 1895 by Borsig of Berlin for the 3ft 6in-gauge Oslav–Volgoda–Archangelsk railway in northern Russia. The design was chosen simply because it allowed a large boiler with good steaming qualities to be allied with a light axle loading and 100 per cent adhesive weight. Engines of this type made in the early 1900s by the Putilov works had 43in-diameter coupled wheels, a boiler pressure of 171psi, and cylinder diameters of 13.8in (high) and 17.7in (low), the common stroke measuring 22in. In working order, with a full load of coal and water, engines and tenders weighed 46.2 tons and 23.6 tons respectively.

The first large Mallet to be made in the USA dated from 1903, when one machine was ordered by the Baltimore & Ohio Railroad from the American Locomotive Company of Schenectady. Exhibited at the St Louis World's Fair of 1904, the engine was a four-cylinder compound with two 20 × 32in high-pressure cylinders, two 32 × 32in low-pressure cylinders and a weight of 212½ tons in working order.

Intended to haul large freight trains, it was successful enough in service but not suited to fast running. Leading trucks were soon added to the basic design to improve lateral stability on curves.

The first Mallets to be made in Britain were four 0-6-6-0 compounds built in Glasgow by the North British Locomotive Co. Ltd. Supplied in 1907–09 to the Peking–Kalgan Railway in China, they had two 18 × 28in high- and two 28¾ × 28in low-pressure cylinders. The diameter of the driving wheels was 4ft 3in; boiler pressure was 225psi.

OOO+OOO

0-6-0+0-6-0, or C+C

Alternative European classifications: 030+030 or 6/6.
Only two Garratts of this layout were ever made by Beyer, Peacock & Co. They were sent from Britain in 1913 to the 2ft 6in-gauge Arakan Flotilla Railway in Burma.

OOOOOOo

0-12-2, or F1

Alternative European classifications: 061 or 6/7.
This particular wheel arrangement was confined largely to heavy freight engines. The only others to be built in this form were two rack and adhesion 24-Class examples, built for the General Manuel Belgrano Railway in southern Argentina by Metallwarenfabrik Esslingen as late as 1955. These had two external cylinders for the adhesion engine, and two smaller auxiliary cylinders for the rack-drive mechanism. Unfortunately, made at a time when interest in steam was declining, their service life was short.

OOO=OOOo

0-6=6-2, or CC1

Alternative European classifications: 331, 6/7.
Locomotives of this notation were running in the USA prior to 1914. They are assumed to have been 0-6-6-0 bankers or 'pushers' fitted with wide fireboxes in an attempt to improve their steaming qualities. This would have necessitated a trailing truck to provide extra support for the rearward overhang of the frames.

OOOo:oOOO

0-6-2:2-6-0 or C11C

Alternative European classifications: 3113, 6/8.
This layout was unique to the du Bousquet heavy haulage tank engines, which were not only compounds but also articulated. The four cylinders were placed at the inner ends of the power units to minimise problems with leaking steam pipes. The locomotives proved to be powerful and free running. The Nord and Est railways bought sixty-one of them, but all but fourteen were transferred in 1921 to the 'Grande Ceinture' serving all the major Paris railway terminals. The Ceinture subsequently acquired thirty-eight engines of a slightly different (and more powerful) type; a few were used in China, and there were nineteen on the 5ft 6in-gauge Andalusian railway in Spain. The last French du Bousquets was withdrawn in 1952.

oOOOOOO

2-12-0, or 1F2

Alternative European classifications: 160 or 6/7.
Popular name: Centipede (nickname only).
Locomotives with this wheel arrangement were restricted invariably to slow speed or ultra heavy services. The first 2-12-0 was a unique four-cylinder compound 100.01, built in 1911 by the Austrian state railway factory in Floridsdorf to the design of Karl Gölsdorf. It was intended to work trains of 300–350 tonnes over the lightly laid Arlberg line, which necessitated an axle loading of less than 14 tonnes.

The cylinder dimensions were 45 × 68cm (high pressure) and 76 × 68cm (low pressure), boiler pressure being 16at (235psi). Driving wheels were 1.45m; weight in running order was 141.5 tonnes.

The only 2-12-0 to have been made in quantity was a four-cylinder compound designed by Eugen Kittel for the Royal Württemberg State Railways, forty-four being delivered by Metallwarenfabrik Esslingen in 1917–24. Designed for the Geislinger Sieg, 6.5km of the Stuttgart–Ulm route graded at more than 2 per cent, the engines had 50 × 65cm (high-pressure) and 75 × 65cm (low-pressure) cylinders; coupled wheels measured 1.35m, boiler pressure was 14.5at (213psi) and weight totalled 154 tonnes in running order. The last survivors were withdrawn from Austria in 1957.

The last 2-12-0 was the six-cylinder 160A.1 designed by André Chapelon made in 1947 by SNCF workshops. It had two 52 x 54cm cylinders, four of 64 × 65cm, a boiler pressure of 17.5at (257psi) and a weight of 217.2 tonnes.

oOOO=OOO

2-6=6-0, or 1CC

Alternative European classifications: 1330, 6/7.
The first example, a tank engine, was built in 1906 in New Zealand for the Rimutaka Incline. Tender engines were built in quantity by Alco, the North British Locomotive Company, Maffei and the Montreal Locomotive Works for the 3ft 6in-gauge Natal and South African Railways (Classes MA-MC, MC1, MJ and MJ1, 1909–21).

The British-built examples had cylinder diameters of 17½in (high pressure) and 28in (low pressure), the stroke being 26in. The coupled wheels had a diameter of 3ft 10in, boiler pressure was 200psi, and weights in working order were 95.5 tons (engine) plus 43.4 tons (tender). A few 2-6-6-0 Mallets also ran on the Denver, North Western & Pacific Railroad in the USA.

oOOOOOOo

2-12-2, or 1F1

Alternative European classifications: 161 or 6/8.
This particular wheel arrangement was confined largely to heavy freight engines. The first representative of this genre was the pioneer tank engine of the 61 Class, twenty-four of which were built for the 3ft 6in-gauge Javanese railways by Hannoversche Maschinenfabrik AG (22 in 1912–13) and Werkspoor (two in 1920). They had two external cylinders, 54 × 51cm, boilers pressed to 12at (176psi) and 1.1m wheels. They weighed about 75 tonnes in working order and were 13.08m long. The only others were two 2-12-2T rack-and-adhesion locomotives built in the Austrian state engineering factory in Floridsdorf in 1941. The cylinders measured 61 × 57cm (adhesion) and 40 × 50cm (rack engine); the boiler pressure was 15.5at (228psi). Driving wheels were 1.05m and the engines weighed 125.9 tonnes in running order.

oOOO=OOOo

2-6=6-2, or 1CC1

Alternative European classifications: 1331, 6/8.
This was basically a 2-6=6-0 with the frames extended rearward to carry a large firebox. The earliest was a compound design introduced on the Great Northern Railroad in 1906. Made by the Baldwin Locomotive Company of Philadelphia, the engines had cylinders measuring 22 × 32in (high pressure) and 33 × 32in (low pressure). The diameter of the coupled wheels was 4ft 7in and the engines weighed about 224.4 tons in working order.

Mallets of 2-6=6-2 type were also made by converting existing engines, among the earliest exponents being the Chicago Great Western and Atchison, Topeka & Santa Fe railroads. The Chicago locomotives had new Baldwin-made chassis extensions, but the AT&SF examples were each converted from two Vauclain-type compound 2-6-2s. The two-engine conversions had a new superheater, reheater and feedwater heater above the front chassis, the boiler sections being joined by a complicated ball and socket joint to avoid excessive lateral movement on curves. They proved to be poor steamers at speed.

Simple expansion locomotives were made by Baldwin in 1930 for the Baltimore & Ohio Railroad, their 5ft 10in driving wheels suiting them to fast freight trains and even passenger services. Among the last 2-6=6-2s were the H 6 compounds, built by the Baldwin Locomotive Company for the Chesapeake & Ohio Railroad in 1949; the last steam engines to be made by Baldwin for service in North America.

These massive machines had 4ft 8in coupled wheels, driven by two 22 × 32in high-pressure (back) and two 35 × 32in low-pressure (front) cylinders. Tractive effort was 98,700lb, when operating as a simple expansion machine; overall length was 96in 9¾in, weight being 322 short tons. Mallets of this type were used in Serbia, on 2ft-gauge-track, and in South Africa where a few were built by Alco and NBL for 3ft 6in-gauge (classes MD, MF and MH, 1910–6); a single example of the P34 class was made in the USSR.

oOOO+OOOo

2-6-0+0-6-2, or 1C+C1

Alternative European classifications: 130+031 or 6/8.

The first six of these were made for the Western Australia government railway in 1911, for 3ft 6in-gauge track. Thirty others were delivered to the London, Midland & Scottish Railway in 1927–30 to haul the Toton–Cricklewood coal trains. Made by Beyer, Peacock & Co. Ltd they had 5ft 3in driving wheels, four 18½ × 26in cylinders, boilers pressed to 190psi, and weighed 152.5 tons empty. Overall length was 87ft 10½in. These Garratts were not particularly successful, partly because the enforced use of standard LMS axleboxes, which were much too flimsy, promoted overheating. The longest-lived representative eventually went to the scrapyard in 1956.

The last to be made was a single 3ft 6in-gauge Garratt supplied in 1939 by Beyer, Peacock & Co. Ltd to the Portland Cement Company of Australia.

An LMS 2-6-0+0-6-2 Garratt. Made by Beyer, Peacock & Co. Ltd in 1930, No. 4983, subsequently 7983 and then 47983, displays the self-trimming bunker that was subsequently fitted to earlier locomotives.

oOOOOOOoo

2-12-4, or 1F2

Alternative European classifications: 162 or 6/9.
This particular wheel arrangement was confined to heavy freight engines. It was unique to a handful of tank engines built in 1931–43 for the Bulgarian State Railways by Cegielski (about sixteen two-cylinder examples) and Berliner Maschinenbau AG vorm. L. Schwartzkopff (eight three-cylinder examples). The Cegielski locomotives had 70 × 70cm cylinders, boiler pressures of 15.5at (228psi) and driving wheel diameters of 1.34m. Weight in running order was a staggering 149.7 tonnes.

oOOO=OOOoo

2-6=6-4, or 1CC2

Alternative European classifications: 1332, 6/9.
This was little more than a 2-6=6-2 Mallet with a trailing bogie to provide greater support for an enlarged 'Super Power' firebox pioneered by Lima Locomotive Works. The earliest was built in 1934 for the Pittsburgh & West Virginia Railway, followed by ten 'fast freight' engines for the Seaboard Air Line, by Baldwin, and a substantial number built in Roanoke by the Norfolk & Western. Introduced on the Norfolk & Western Railroad in 1936, built in the company's own workshops, the last of these Class-A compound Mallets was completed in 1950. They were specifically designed to achieve quite high speeds, the diameter of the driving wheels being 5ft 10in. Thus they were rated capable of 78mph on passenger work. The four cylinders measured 24 × 30in, tractive effort being 125,900lb. The locomotives were 121ft 9¼in overall and weighed about 471 short tons in working order. Locomotives of this notation were rarely encountered elsewhere.

oOOOo+oOOOo

2-6-2+2-6-2, or 1C1+1C1

Alternative European classifications: 131+131 or 6/10.
First made in 1912, when two 3ft 6in-gauge examples were supplied by Beyer, Peacock & Company to the Tasmanian government, this was one of the most popular Garratt layouts available prior to 1925 – though subsequently replaced by much larger locomotives. Orders were accepted by Beyer, Peacock & Co. Ltd from South Africa (1923 onward) and India, from 1925. The last examples were seven made for the 2ft-gauge railways of Tsumeb Corporation of South Africa in the early 1950s.

A Seaboard Airlines 2-6=6-4 R1-Class Mallet, supplied by Baldwin in 1935. The identical cylinders show that this is a simple expansion design.

oOOO=OOOooo

2-6=6-6, or 1CC3

Alternative European classifications: 1333, 6/10.
Popular name: Allegheny.
The first of sixty Chesapeake & Ohio Railroad-Class H 8 Mallets was built by the Lima Locomotive Company in 1941, the last being delivered in 1944. Powered by four 22½ × 33in cylinders, with 5ft 7in driving wheels, boiler pressure of 260psi and a nominal tractive effort of 110,200lb, they were capable of hauling 160-wagon coal trains in the Alleghenies – loads of 11,500 tons – owing to the steam-raising capabilities of the boiler and its extra large firebox.

Each locomotive was 125ft 8in long and weighed about 389 short tons without its tender. The wheel arrangement was confined to the Chesapeake & Ohio and Virginian railroads.

ooOOOOOOo

4-12-2, or 2F1

Alternative European classifications: 261 or 6/9.
Popular name: Centipede.
This particular wheel arrangement was unique to a single class of super-heavy freight engine. Except for the unsuccessful Soviet 4-14-4 (q.v.), they were the largest of all rigid frame designs. The notation was restricted to eighty-eight three-cylinder representatives of the 9000 Class, built in 1926–30 for the Union Pacific Railroad by the American Locomotive Company of Schenectady. They had two 27 × 32in outside and one 27 × 31in inside cylinders with Gresley-type conjugated valve gear, but were not entirely successful. Intended to pull fast freight trains in flatlands, the fifth and sixth axles were allowed lateral movement. The locomotives had a boiler pressure of 230psi, giving a tractive effort of 97,664lb; they were 102ft 7in long and weighed about 391 short tons with the tender.

The Union Pacific 4-12-2 *Centipede*, the longest of all rigid wheelbase locomotives with the exception of the Soviet 4-14-4. A total of eighty-eight of the type were supplied by Alco in 1926–30 in five sub-variants.

ooOOO=OOOoo

4-6=6-4, or 2CC2

Alternative European classifications: 2332, 6/10.
Popular name: Challenger.

The first fifteen of these, Class CSA-1, were built by the American Locomotive Company in 1936 for the Union Pacific Railroad. They had 5ft 9in driving wheels, which gave a respectable turn of speed in addition to excellent hauling capacity. However, Challengers had a habit of slipping if power was applied too quickly. They were confined to the Union Pacific and Denver & Rio Grande Western roads. The perfected Union Pacific 3900 Class had four 21 × 32in cylinders and a nominal tractive effort of 97,350lb; total weight in working order was 535 short tons, and overall length was 121ft 11in.

ooOOO+OOOoo

4-6-0+0-6-4, or 2C+C2

Alternative European classifications: 230+032 or 6/10.

Five Garratts of this type were supplied by Beyer, Peacock & Co. Ltd to the metre-gauge Mogyana Railway (India) in 1912–14. In addition, the articulated *River Mite* of the 15in-gauge Ravenglass & Eskdale Railway – not a true Garratt! – was made in the Ravenglass workshops in 1928 by fitting a new boiler obtained from the Yorkshire Engine Company to frames made by cannibalising obsolescent locomotives. Unfortunately, the boiler mounting proved to be troublesome in service and so, in 1938, *River Mite* was withdrawn. It was eventually sent for scrap during the Second World War.

Union Pacific Challenger-Class 4-6=6-4 Mallet No. 3901 was the second of the CSA-1 Class, built by Alco in 1936. By 1944, 111 Challengers had been supplied in five sub-variants. The CSA-1 had 5ft 9in driving wheels, four 22 × 32in cylinders and a boiler pressure of 255psi; the locomotive alone weighed about 283 short tons.

ooOOOo+oOOOoo

4-6-2+2-6-4, or 2C1+1C2

Alternative European classifications: 231+132 or 6/12.

Five Garratts of this type were uncommon. The earliest were three 3ft 6in-gauge examples sold by Beyer, Peacock & Co. to the New Zealand government in 1928. It is sometimes claimed that the first 'Double Pacific' Garratts were sent to Brazil in 1927, but these were built as 2-6-2+2-6-2 large-wheel 'express' engines for the São Paulo Railway in Brazil. The São Paulo locomotives were subsequently converted by the substitution of leading bogies so that more water could be carried; the changes were made in Brazil from *c.* 1934 onward. The last examples were six 'Double Pacifics' supplied to the 3ft 6in-gauge Nigerian railway in 1943.

ooOOOoo+ooOOOoo

4-6-4+4-6-4, or 2C2+2C2

Alternative European classifications: 232+232 or 6/14.
This notation was the most popular of all Garratts. The first were four purchased on behalf of the Sudanese government railway in 1936–37. These were not successful in the sandy conditions that prevailed in the Sudan and were sold to Rhodesia for service on 3ft 6in-gauge track. The Rhodesian Railways thereafter acquired seventy similar locomotives (15 Class) in six batches, delivered between 1941 and 1952. The last batch was made under licence by Société Franco-Belge.

Seven Driving Axles

OOOOOOO

0-14-0, or G

Alternative European classifications: 070, 7/7.
This notation seems to have been unique to the original version of *Bavaria*, a multi-drive machine supplied by J.A. Maffei of Munich to the Semmering locomotive trials in 1851. Although a glance suggested that the machine could be an 0-4-4-6-0, everything was connected. Two 20 × 30in outside cylinders, supplied with steam at 125lb/in², drove directly on to the fourth axle, which was coupled to the third axle by conventional side rods and to the second driven axle by a chain between the second and third. The front two axles, connected together, were combined in a bogie to allow flexibility in the wheelbase. The wheels beneath the 'tender', coupled externally by side rods, were also connected by chains between the fourth and fifth axles. The locomotive weighed 71.6 tons, making it by far the heaviest of its day. However, though *Bavaria* rather surprisingly won the trials, complexity and persistent breakages ensured that the locomotive was scrapped after occasionally running as an 0-8-6 (when the chain-link between the fourth and fifth axles broke) or even as a 4-4-6 when both chain drives broke! The chains had soon been replaced with gears, but even these failed to cure the inherent problems.

oOOO=OOOO

2-6=8-0, or 1CD

Alternative European classifications: 1340, 7/8.
The first of these Mallets was made for the Great Northern Railroad by adding a new six-driver Baldwin-made chassis extension to a 2-8-0 *Consolidation* conversion. The original 20 × 32in cylinders formed the high-pressure system, the low-pressure units being carried on the new section. The results were not altogether successful.

ooOOOOOOOoo

4-14-4, or 2G2

Alternative European classifications: 272 or 7/11.
This particular wheel arrangement was confined to a single super-heavy freight engine. The locomotive, which was being constructed by Krupp as a 2-14-4, finally emanated from the Lugansk locomotive factory in the USSR in December 1934. Bearing the Class name 'A. Andreev' and intended to haul massive 2,750-tonne coal trains up 1-in-100 gradients at 25kph, it weighed 333 tonnes with its tender and was 33.74m (110ft 8in) long. The two outside cylinders measured 74 × 81cm (29⅛ × 31⅞in) and the driving wheels had a diameter of 1.6m (5ft 4in). However, even though the third, fourth and fifth driven axles had flangeless wheels, the wheelbase was much too rigid; the engine was totally unsuccessful and was scrapped *c.* 1960 after lying out of service for many years.

The Soviet 4-14-4, known as 'Andrey Andreev' in recognition of its sponsor, was the world's largest rigid frame locomotive. It was ordered from Krupp in the 1930s but, incomplete, was taken back to Lugansk and completed with a leading bogie. Propaganda trips undertaken in 1935 were disastrous. Even though the third, fourth and fifth axles had flangeless wheels, and the second and sixth axles were allowed radial movement, AA 20-1 wrecked the track. Rapidly withdrawn into store in Shcherbinka, the massive locomotive was quietly scrapped.

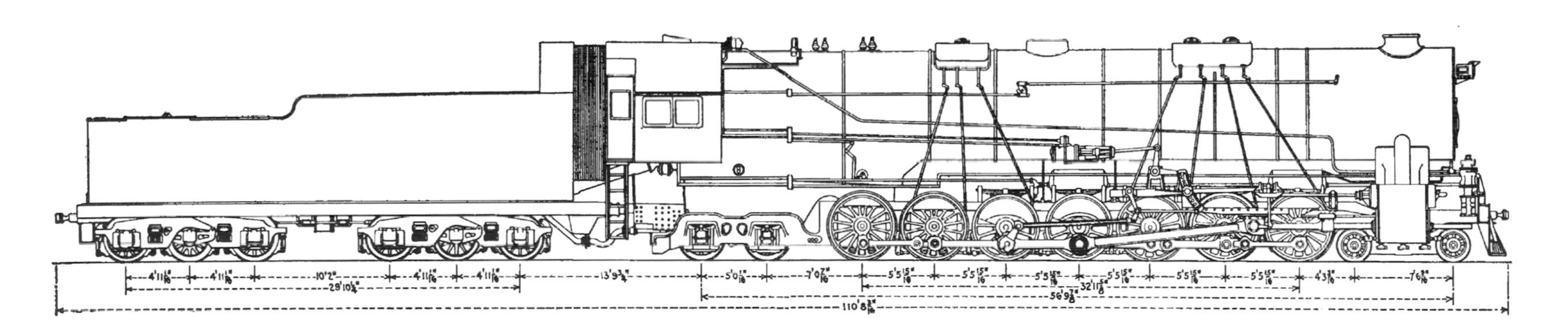

Eight Driving Axles

OOOO=OOOO

0-8=8-0, or DD

Alternative European classifications: 0440, 8/8.

The first Baldwin-made Mallets of this type were delivered to the Erie Railroad in 1907 to haul 3,000-ton coal trains over a short section of steeply graded track. The representatives of Class L-1 were the largest of all Camelback or Mother Hubbard designs, with the driver's cab midway along the boiler; weight in running order was about 275 short tons. Diameters of the high- and low-pressure cylinders were 25in and 39in respectively, the common stroke being 28in; boiler pressure was 215psi, giving a tractive effort of 88,890lb. The diameter of the coupled wheels was merely 4ft 3in, befitting a heavy freight role. However, the L-1 Mallets were rebuilt in 1921 in 2-8=8-2 form but had gone by 1930.

oOOOO=OOOO

2-8=8-0, or 1DD

Alternative European classifications: 1440, 8/9.

Though rapidly superseded by the more effectual 2-8=8-2, Mallets of this type ran on Great Northern, Union Pacific, Baltimore & Ohio and a few other railroads. The earliest were four short-lived 2-8-0 conversions made in 1911 by the Atchison, Topeka and Santa Fé.

oOOOO=OOOOo

2-8=8-2, or 1DD1

Alternative European classifications: 1441 or 8/10.

The first of this type was built in 1909 by Baldwin for the Southern Railroad. A typical 1911-vintage Mallet for the Virginian Railroad had cylinders measuring 28 × 32in (high pressure) and 44 × 32in (low pressure). The diameter of the coupled wheels was 4ft 8in and the locomotive weighed 283.5 tons in running order. These Mallets were usually progressions from the 2-8-8-0, with the frames extended rearwards to support a larger firebox. The Norfolk & Western workshops built the last of the 2-8-8-2 Y6b compounds in 1952, with two cylinders measuring 25 × 32in and two of 39 × 32in. Driving wheel diameter was 4ft 10in, tractive effort being 150,206lb if high-pressure steam was admitted to all four cylinders simultaneously. Overall length was 114ft 10½in, weight in working order being 445 short tons.

This Seaboard Airlines 2-8=8-2 Class-A Mallet was built by the Richmond Locomotive Works (part of the Alco group) in 1918.

oOOOO+OOOOo

2-8-0+0-8-2, or 1D+D1

Alternative European classifications: 14+41 or 8/10.
The first of these Garratts was supplied to the Burma Railway in 1924, but the best known example – also the largest ever locomotive to run in Britain – was the solitary No. 2925 made in 1925 by Beyer, Peacock & Co. Ltd (works No. 6209) for the London & North Eastern Railway to assist goods trains on the bank between Penistone and Wath. It had two sets of three 18½ × 26in cylinders, 4ft 8in-diameter wheels, and a total wheelbase of 79ft 1in. Weight in running order was 178 tons.

The last representatives were ten metre-gauge locomotives supplied to the War Department in 1943, destined to serve in Burma.

oOOOO:OOOOoo

2-8=8-4, or 1DD2

Alternative European classifications: 1442 or 8/11.
Popular name: Yellowstone.
Introduced on the Northern Pacific in 1928, when the first Alco-made Z-1 was delivered, locomotives of this type (designated M3 and M4) were also used by the Duluth, Missabe & Iron Range Railroad, and by the Baltimore & Ohio Railroad – where the first of the Baldwin-made EM-1-Class appeared in 1944. The inclusion of a trailing bogie enabled a large 'Super Power' firebox to be fitted. The eighteen examples of DM&IRR Classes M3 and M4, made by Baldwin in 1940–43, had four 26 × 32in cylinders, 5ft 3in coupled wheels and a tractive effort of about 140,000lb; they could move loads of 15,000 short tons, maximum speed being 30mph. The engines were 127ft 8in long and weighed 569 short tons in working order. The first Southern Pacific cab-forward locomotives took this form, as did a large number built by the Baltimore & Ohio during the Second World War.

oOOOOo+OOOO

2-8-2+0-8-0, or 1D1+D

Alternative European classifications: 141+040 or 8/10.
This notation seems to have been confined to *River Esk*, built in 1924 as a 2-8-2 by Davey, Paxman & Co. Ltd for the 15in-gauge Ravenglass & Eskdale Railway. The engine was rebuilt in 1927, gaining an eight-wheel booster forming the tender frame, but experience showed that the boiler did not raise enough steam to supply the cylinders when the train was climbing what are quite severe gradients. *River Esk* had reverted to 2-8-2 form by 1939.

oOOOOo+oOOOOo

2-8-2+2-8-2, or 1D1+1D1

Alternative European classifications: 141+141 or 8/12.
Garratts of this type were less common than some of the other eight-coupled patterns. The first three were supplied to the so-called 'Nitrate railways' in Chile in 1926, and the last were Class-16A examples supplied by Beyer, Peacock & Co. Ltd to the Rhodesian Railways in 1953–54. (Ten South Australian 400-Class locomotives ordered later in 1953 were apparently delivered from licensees Société Franco-Belge before the last 16A had been dispatched from Manchester.)

ooOOOO=OOOOo

4-8=8-2, or 2DD1

Alternative European classifications: 2441 or 8/11.
The Southern Pacific Railroad had an unusual number of tunnels and snow-sheds, which affected crewmen to the point where primitive breathing equipment had to be used to avoid smoke poisoning. The AC or 'Articulated Consolidations' series of large Mallets ordered in 1928 had the cab at the leading end, with the boiler reversed so that the chimney was directly ahead of the tender. Except for the coal-fired AC-9 group made by Lima in 1939, which had a conventional layout, the locomotives were oil-fired. No fewer than 145 were delivered by Baldwin, beginning in 1930.

A typical AC 12 had four 24 × 32in cylinders, 5ft 3in coupled wheels and a nominal tractive effort of 124,300lb. It was 111ft 9in long and weighed 526 short tons.

ooOOOO=OOOOoo

4-8=8-4, or 2DD2

Alternative European classifications: 2442 or 8/12.
Popular name: Big Boy.
The name arose from an unofficial nickname chalked by an Alco employee on the smokebox door of the first to be built of this type, the longest of all the articulated two-unit Mallets to run in North America. Capable of

The Union Pacific 4-8-8-4 *Big Boy*, seen at Laramie, Wyoming, in August 1957, was legendary. No. 4008 of the first group, supplied by Alco in 1941, weighed 345 short tons without its tender. (Dr Richard Leonard; photographed by David Leonard)

hauling vast loads over stupendous distances, they served with distinction until displaced by multi-unit diesels in the 1950s. The first of the initial order for twenty locomotives was completed in 1940 by the American Locomotive Company ('Alco') of Schenectady, New York, for the Union Pacific Railroad. Despite their immense size, the *Big Boys* had 5ft 8in coupled wheels and were rated at speeds of up to 68mph for passenger or fast freight work. They had four 23¼ × 32in cylinders and a nominal tractive effort – an indication of their power – of 135,375lb. This was more than seventy times as great as the tractive effort of *Locomotion No. 1*, built 125 years earlier. The five engines of the 1944 series, which were slightly larger than their predecessors, were 132ft 10in long and weighed a staggering 604 short tons in running order.

ooOOOO+OOOOoo

4-8-0+0-8-4, or 2D+D2

Alternative European classifications: 24+42 or 8/12.

The only Garratts of this type were unique to the Bengal & Nagpur Railway. Beyer, Peacock & Co. Ltd supplied twenty-six of them – twenty with Walschaerts', three with Caprotti and three with Lentz gear – in 1930–31. it is a pity none of the trial records have been found, as the differences between valve gears, if any, would have been instructive.

ooOOOOo+oOOOOoo

4–8–2+2–8–4, or 2D1+1D2

Alternative European classifications: 241+142 or 8/14.

The first batch of many Garratts made to this popular layout were four of the 50- (later EC1-) Class examples supplied to the metre-gauge Kenya & Uganda Railway in 1926. Six 3ft 6in-gauge examples, ordered at much the same time, were sent to the Benguela railway in Mozambique. The last twelve were sent to the Buenos Ayres Great Southern Railway in 1928.

A 4-8-2+2-8-4 Garratt of the GO class, twenty-five of which were supplied to the South African Railways in 1954 by Henschel. They were a lighter form of the GM series, intended to be used on lightly laid branch lines.

ooOOOOoo+ooOOOOoo

4-8-4+4-8-4, or 2D2+2D2

Alternative European classifications: 242+242 or 8/16.

The first of these, the largest of all Garratts, was a batch of six supplied to the metre-gauge Kenya & Uganda Railway by Beyer Peacock in 1939, a few months before the start of the Second World War. Classified as 57 (later EC3), engines of this type had 4ft 6in-diameter driving wheels. The last of the type were ordered by the New South Wales government railways in 1956. The engines were successful enough to remain in service for years; one from an earlier group is preserved in working order.

4-8-4+4-8-4 Garratt No. 6029 was one of a group supplied in 1953 to the New South Wales railways by Beyer, Peacock & Co Ltd. Now named *City of Canberra*, No. 6029 was originally intended to pull heavy coal trains in Hunter Valley. (Canberra Railway Museum)

Ten or More Driving Axles

oOOOOO=OOOOOo

2-10=10-2, 1EE1

Alternative European classifications: 1551 or 10/12.

The first five engines of this pattern were made for the Atchison, Topeka & Santa Fe Railroad prior to 1914. They were converted from 2-10-2 tandem compound freight locomotives, one supplying the cab, the rear set of running gear and the low-pressure portion of the cylinders sleeved down to a diameter of 28in; the other provided the front set of running gear. However, the need for an entirely new boiler shell and new low-pressure cylinders (with a diameter of 38in) made work uneconomic. Unfortunately, the engines could not maintain steam pressure if speed rose above 15mph and they were soon replaced by new machines with much larger boilers.

More successful were the colossal engines built by the American Locomotive Company of Schenectady for the Virginian Railroad. Used to haul iron-ore trains, these compounds had two high-pressure cylinders measuring 30 × 32in, two low-pressure units of 48 × 32in, and 4ft 8in coupled wheels.

Though boiler pressure was comparatively low – 215psi – simple expansion tractive effort was 176,600lb. Weight in working order was 449 short tons, owing to the comparatively compact tender. The engine was 73ft overall, including the pilot but excluding the rearward extension of the cab roof.

The 2-10=10-2 Atchison, Topeka & Santa Fe Railroad 3000-class Mallet of 1910 was a step too far! The boiler was far too long to be effective, particularly as only about half its length was given over to steam raising. No. 3003 is shown at Winslow, Arizona, *c.* 1913. (Drawing taken from *Engineering*, 10 July 1914)

oOOOO=OOOO=OOOOo

2-8=8=8-2, or 1DDD1

Alternative European classifications: 14441 or 12/14.
Four of these extraordinary Triplex Mallets, based on patents granted to George Henderson of the Baldwin Locomotive Company, were made for the Erie Railroad. The first, *Matt H. Shay* (pictured on page 55), was delivered in 1913. The engine had six 36 × 32in cylinders, the central high-pressure pair exhausting to the low-pressure pairs. The coupled wheels had a diameter of 5ft 3in, which, with a boiler pressure of 210psi, gave a tractive effort of about 160,000lb; the engine was 103ft 4in long (excluding couplers) and weighed 427 short tons in running order. Used for banking and heavy freight haulage, these Triplex Mallets were not especially successful. Though capable of moving prodigious loads at very slow speeds, they suffered from insufficient steam-raising capacity if speed was raised beyond 15mph.

oOOOO:OOOO:OOOOoo

2-8=8=8-4, or 1DDD2

Alternative European classifications: 14442 or 12/15.
Only one of these huge machines was ever made by the Baldwin Locomotive Company of Philadelphia. No. 700 was delivered to the Virginian Railway in 1919 for use as a 'pusher' on ultra-heavy coal trains. Essentially similar to *Matt H. Shay* (above), the locomotive had six 34 × 32in cylinders, 4ft 8in coupled wheels and a tractive effort of 166,300lb. The engine was 103ft 7in overall, excluding couplers, and weighed 422 short tons. It was not successful, owing to continual steam leaks, and was reconstructed *c.* 1922 as a conventional 2-8=8-0. The power tender formed the basis for a new 2-8-4.

BIBLIOGRAPHY

Ahrons, Ernest L.: *The British Steam Locomotive, 1825–1925*. The Locomotive Publishing Co., London, 1926 (but since reprinted several times).

Anonymous or Unattributed: 'Compound Locomotive: London and North-Western Railway' [Webb type]. *Engineering*, May 1885.

—'Compound Express Locomotive for the North-Eastern Railway' [Worsdell type]. *Engineering*, March and April 1888.

—'Compound Express Locomotive for the Northern Railway of France' [du Bousquet type]. *Engineering*, June 1898.

—'Compound Goods Locomotive for the North-Eastern Railway' [Worsdell type]. *Engineering*, September 1887.

—'Compound Goods Locomotive: Paris Exhibition' [Sauvage type]. *Engineering*, December 1889.

—'Compound Passenger Locomotive for the North-Eastern Railway' [Worsdell type]. *Engineering*, June 1887.

—'Early Compounds of the P.L.M. Railway'. *The Locomotive, Railway Carriage and Wagon Review*, The Locomotive Publishing Co., London, in five parts, January–August 1935 (vol. XLI).

—'Express Compound Locomotive with Auxiliary Gear' [Krauss type]. *Engineering*, May 1897.

—'Express Locomotive, Krauss System'. *Engineering*, April 1901.

—'Four-Wheel Coupled Three Cylinder Compound Locomotive' [Johnson/Smith type]. *Engineering*, May 1900.

—'Messrs. Schneider & Co.'s [locomotive] Works at Creusot'. *Engineering*, September 1898.

—'The Webb System of Compounding Locomotives'. *Engineering*, May 1894.

—'Three-Cylinder Compound Locomotive for the North-Eastern Railway' [Smith type]. *Engineering*, July 1901.

Burton, Anthony: *The Railway Empire*. John Murray (Publishers) Ltd, London, 1994.

Clark, Danniel Kinnear, MInstCE, MIME: *The Steam Locomotive* ('A Treatise on Steam Engines and Boilers …'). Blackie & Son Ltd, London, etc., 1891.

Davies, Harold: *North American Steam Locomotive Builders & Their Insignia*. TLC Publishing, Inc., USA, 2005.

Ferneyhough, Frank: *Liverpool & Manchester Railway, 1830–1980*. Robert Hale, London, 1980.

Gairns, J.F.: *Locomotive Compounding and Superheating* ('A Practical Text-Book for the use of Railway and Locomotive Engineers, Students, and Draughtsmen'). Charles Griffin & Co. Ltd, London, 1907.

Jamieson, Andrew, CE: *A Text-Book on Steam and Steam Engines* ('Specially arranged for the use of Science and Art …'). Charles Griffin & Co., Ltd, London, 1892.

Johnson, Samuel W.: Presidential Address to the Institute of Mechanical Engineers ('On Our Railways'). *Engineering*, four parts, May–June 1898.

Jones, Robin: *Steam's New Dawn* ('Britain's Third Century of Steam Locomotives'). Halsgrove, Wellington, Somerset, 2011.

—*The Rocket Men* ('George and Robert Stephenson. The Men who re-shaped the World'). Mortons Media Group, Horncastle, Lincoln, 2013.

Lowe, James W., CEng, FIMechE: *British Steam Locomotive Builders*. TEE Publishing, Hinckley, Leicestershire, 1975 (a supplement was published in 1984).

McConnell, J.E., CE MInstCE MInstME, and W.J. Maquorn Rankine CE LLD FRS: '6. Locomotive Engines' in *A Record of the International Exhibition, 1862*. Spon, London, 1862.

Mallet, Anatole: 'Compound Articulated Locomotives' (read before the Institution of Mechanical Engineers). *Engineering*, July 1914.

Newcomen Society for the Study of Science and Technology: Dewhurst, P.C., MICE MIMechE: 'The Crampton Locomotive in England' ('Part I. The Outside Cylinder Rear-Driver Pattern' and 'Part II'), *Transactions*, vol. XXX, 1955/6, and 1956/7.

—Dewhurst, P.C., MICE MIMechE: '... "Norris" Locomotives in England' ('Part I. Engines constructed by Norris in Philadelphia, U.S.A.', and 'Part II. Norris-pattern engines constructed by English firms'). *Transactions*, vol. XXVI, 1947/8 and 1948/9.

—Dewhurst, P.C., MICE MIMechE: 'The Fairlie Locomotive' ('Part I. The formative period' and, with Harold Holcroft CEng MIMechE, 'Part II. Later Designs and Productions'). *Transactions*, vols XXXIV, 1962, and XXXVIII, 1966.

—Hills, Richard L., MA DIC: 'Some Contributions to Locomotive Development by Beyer, Peacock & Co'. *Transactions*, vol. XXXX, 1968.

Nock, O.S., BSc CEng FICE, FIMechE MILocoE MIRSE: *The British Steam Locomotive, 1925–1965*. Ian Allan Ltd, London, 1966.

Reed, Brian: 'German Austerities', in *Locomotive Profiles*, vol. 1, 1977.

Riemsdijk, J.T. van: *Compound Locomotives. An International Survey*. Atlantic Transport Publishers, Penryn, Cornwall, 1994.

Rolt, L.T.C.: *Victorian Engineering*. The History Press, Stroud, Gloucestershire, 2010.

Sauvage, Édouard: 'Recent Locomotive Practice in France' (read to the Institute of Mechanical Engineers). *Engineering*, June 1900.

—'Compound Locomotives in France' (read before the Institution of Mechanical Engineers). *Engineering*, March and April 1904.

—'Recent Development of Express Locomotives in France' (read before the Institution of Mechanical Engineers in Paris). *Engineering*, July 1914.

Sinclair, Angus: *Development of the Locomotive Engine* ('A History of the Growth of the Locomotive from its Most Elementary Form ...'). Angus Sinclair Publishing Co., New York, 1907.

Smith, Walter M.: 'Express Locomotives' (read before the Institution of Mechanical Engineers). *Engineering*, in two parts, November 1898.

Tanel, Franco: *An Illustrated History of Trains* ('From Steam Locomotives to High-Speed Rail'). David & Charles, Newton Abbot, Devon, 2011.

Thurston, Robert H., AM LLD DEng: *A Manual of the Steam Engine* ('For Engineers and Technical Schools ... Part I. Structure and Theory'). John Wiley & Sons, New York, and Chapman & Hall, London, 1896.

Twining, E.T.: 'British Crampton Locomotives', in *Model Engineer*, twelve parts, March 1953–March 1954.

Westcott, G.F.: *The British Railway Locomotive, 1803–1853*. HMSO, London, 1958.

Winchester, Clarence (ed.): *Railway Wonders of the World*. George Newnes, London, 1937.

ELECTRONIC MEDIA

In addition to the classic printed works, I consulted a variety of electronic sources of information ranging from Wikipedia entries to dedicated websites. Particularly valuable were the websites of Dr Richard Leonard, for its extensive highly pictorial coverage of North American railroading (Richard Leonard's Steam Locomotive Archive, www.railarchive.net); and of Douglas Self (www.douglas-self.com), whose specialties include the 'weirder end' of locomotive design. I also regularly visited Steamindex (www.steamindex.com), meticulously maintained by Kevin Jones, which contains an unbelievable amount of material on virtually every topic I needed to research. It's a particularly good source of biographical details concerning all the locomotive engineers I've ignored!

ILLUSTRATIONS

Unless credited otherwise in the captions, virtually all the illustrations come from material I have acquired over the past thirty years or from books and prints in the British Engineerium's collection. The line drawings and a majority of the photographs are clearly out of copyright and now in public domain, but the origins of others are unclear. If I have unintentionally infringed copyright, suitable acknowledgements can and will be made in future editions. Please contact me by way of the publisher's address given on page 4.